A
PLANTSMAN'S GUIDE TO
PRIMULAS

PHILIP SWINDELLS

A
PLANTSMAN'S GUIDE TO
PRIMULAS

PHILIP SWINDELLS

SERIES EDITOR
ALAN TOOGOOD

WARD LOCK LIMITED · LONDON

ACKNOWLEDGEMENTS

The publishers gratefully acknowledge the following
agencies and photographers for granting permission to
reproduce the following colour photographs: Andrew
Lawson (pp. 8/9); Photos Horticultural Picture Library
(pp. 13, 16/17, 25, 28/29, 33, 37, 40, 45, 65, 69, 81,
88/89, 101 and 116/117); Pat Brindley (pp. 36, 56, 80,
84, 104 and 121); Harry Smith Horticultural
Photographic Collection (pp. 48, 57, 72, 77 and 92);
Tania Midgley (pp. 96/97).

All line drawings are by Nils Solberg.

Text © Ward Lock Ltd 1989
Line drawings © Ward Lock Ltd 1989

First published in Great Britain in 1989
by Ward Lock Limited, 8 Clifford Street
London W1X 1RB, an Egmont Company

House editor Denis Ingram

Text filmset in Times
by Dorchester Typesetting Group
Printed and bound in Great Britain by
Hazel, Watson & Viney Ltd.
Member of the BPCC Group,
Aylesbury, Bucks.

British Library Cataloguing in Publication Data

Swindells, Philip
   A plantsman's guide to primulas
   1. Gardens. Primulas. Cultivation &
   exhibition, – Manuals
   I. Title   II. Series
   635.9'33672

ISBN 0-7063-6739-1

CONTENTS

PUBLISHER'S NOTE

Readers are requested to note that in order to make the text intelligible in both hemispheres, plant flowering times, etc. are described in terms of seasons, not months. The following table provides an approximate 'translation' of seasons into months for the two hemispheres.

Northern Hemisphere		Southern Hemisphere
Mid-winter	= January	= Mid-summer
Late winter	= February	= Late summer
Early spring	= March	= Early autumn
Mid-spring	= April	= Mid-autumn
Late spring	= May	= Late autumn
Early summer	= June	= Early winter
Mid-summer	= July	= Mid-winter
Late summer	= August	= Late winter
Early autumn	= September	= Early spring
Mid-autumn	= October	= Mid-spring
Late autumn	= November	= Late spring
Early winter	= December	= Early summer

Captions for colour photographs on chapter opening pages:

Pp. 8-9 Though they need a degree of care and attention to detail to grow successfully, *Primula allionii* and its myriad named varieties are among the most popular alpine-house primulas.

Pp. 16-17 The candelabra primula, *P. beesiana*, is a vigorous grower, flowering in early summer. Highly recommended for the bog garden or poolside. Grows well in sun or partial shade.

Pp. 28-29 Primula helodoxa is a candelabra species recommended for a moist border or bog garden, provided its roots are not standing in water. It makes a magnificent show during the summer.

Pp. 88-89 A first-class hardy perennial for the peat garden is spring-flowering *Primula whitei*. It grows best when planted in the joints between peat blocks.

Pp. 116-117 The primrose, *Primula vulgaris*, is a widely distributed and much-loved spring-flowering species. An easily grown plant which naturalizes itself in grass or woodland.

EDITOR'S FOREWORD

This unique series takes a completely fresh look at the most popular garden and greenhouse plants.

Written by a team of leading specialists, yet suitable for novice and more experienced gardeners alike, the series considers modern uses of the plants, including refreshing ideas for combining them with other garden or greenhouse plants. This should appeal to the more general gardener who, unlike the specialist, does not want to devote a large part of the garden to a particular plant. Many of the planting schemes and modern uses are beautifully illustrated in colour.

The extensive A-Z lists describe in great detail hundreds of the best varieties and species available today.

For the historically-minded, each book opens with a brief history of the subject up to the present day and, as appropriate, looks at the developments by plant breeders.

The books cover all you need to know about growing and propagating. The former embraces such aspects as suitable sites and soils, planting methods, all-year-round care and how to combat pests, diseases and disorders.

Propagation includes raising plants from seeds and by vegetative means, as appropriate.

For each subject there is a society (sometimes more than one), full details of which round off each book.

The plants that make up this series are very popular and examples can be found in many gardens. However, it is hoped that these books will encourage gardeners to try some of the better, or perhaps more unusual, varieties; ensure some stunning plant associations; and result in the plants being grown well.

CHAPTER ONE

PAST AND PRESENT

Primulas are a cosmopolitan genus of herbaceous and alpine plants, most of them of garden merit. Most of us are familiar with the primroses of hedgerows and polyanthus of cottage gardens. The rather artificial-looking auriculas of the exhibitor are also primulas, as are the giant candelabra varieties of the bog garden and waterside and the tiny mountain species seen in alpine houses. Such a wide diversity of forms, habits and colours should satisfy the most discerning gardener.

No matter what kind of garden you have, whether it is a window box, patio planter, rock garden or extensive borders, there is a primula for you.

The most significant primula for the newcomer to gardening is the polyanthus, that marvellous bright-coloured spring bedding plant loved by gardeners and park directors alike. Its origins are lost in the mists of time, though Parkinson in his famous *Paradisus* (1629) refers to a primrose cowslip which from his description suggests a hybrid between *Primula veris* and *P. vulgaris*, that is, the cowslip and primrose. Later, in 1640, Parkinson described an interesting red-flowered primrose from Turkey which it is thought was used in hybridizing to produce the red polyanthus or big oxlip described by the Worcestershire nurseryman, John Rea, in 1665. He penned eloquent descriptions of the new hybrids of red cowslip or oxlip of that time.

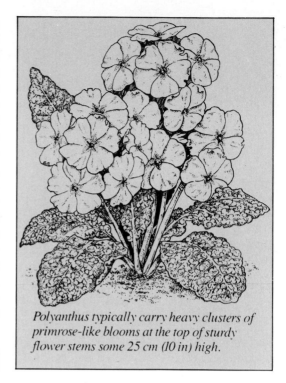

Polyanthus typically carry heavy clusters of primrose-like blooms at the top of sturdy flower stems some 25 cm (10 in) high.

POLYANTHUS

Polyanthus remained largely unchanged for a century or more, when gold edges and lacing were developed round the borders of the petals. It was then taken up as one of the earliest pot plants, becoming a florist's flower rather than a garden plant. It was not until 1880 that the polyanthus we know today emerged. A chance find in the garden of that famous Victorian gardener, Gertrude Jekyll, had a bright yellow polyanthus flower, an ideal parent for the hybrids to be produced from its union with the myriad red and purple varieties then existing.

There were no other colours until 1890, when G. F. Wilson, whose garden at Wisley is now the home of the Royal Horticultural Society, introduced a blue-flowered polyanthus-primrose. The first multicoloured strains of polyanthus were developed from this, Miss Jekyll's Munstead Strain making the greatest impact on home gardeners at that time. The development of this most popular flower has now spread beyond all recognition. Work in the United Kingdom, Germany, the Netherlands and especially the United States, has elevated this plant to heights inconceivable in Parkinson's day.

PRIMROSES

Primroses have a similar history, though it was much later before extensive hy-

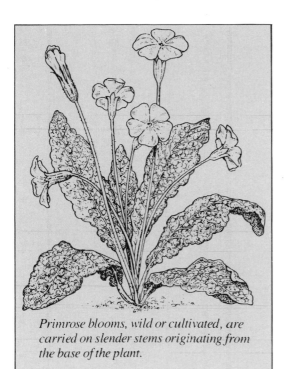

Primrose blooms, wild or cultivated, are carried on slender stems originating from the base of the plant.

bridizing took place. In the 1600s primroses were prized as garden plants and as many different natural variations as possible were collected. These included hose-in-hose and Jack-in-the-green types which were collected just as Victorians gathered together hardy ferns with variously formed fronds.

In the late nineteenth century primroses developed alongside polyanthus, but it was 1913 before the strains of coloured primroses that we know today started to appear. These were largely the work of Arends, a German enthusiast who crossed *P. juliae* with *P. vulgaris* to yield many interesting forms of polyanthus and primrose. Many of his varieties are popular today, but most

need to be propagated vegetatively to ensure they come true to type.

The same applies to the double primroses which were produced in great numbers at much the same time. Cocker Bros. of Aberdeen were one of the most prolific nursery firms, who allegedly raised 137 double seedlings between 1915 and 1918. Known as Bon Accord doubles, they were an important and popular success, but few now remain in cultivation.

AURICULAS

Auriculas enjoyed tremendous popularity during a similar period. Their appeal has waned a little, but they still attract a large group of devotees. The world of the auricula exhibitor is one of mystery to most gardeners, but it is a world of precision and good friendship. The exhibitor of auriculas is one of a special breed who often finds us ordinary gardeners a mystery as we blatantly disregard his system of taxonomic order among his beloved flowers.

For most gardeners an auricula is a garden plant often grown on a rock garden and producing a modest yellow flower, or possibly in shades of purple or pink. These are the auriculas you find in garden centres and very lovely they are too. The enthusiast classifies them as alpine auriculas. But the show and border varieties attract most attention and have the most interesting history.

It is widely accepted that the fancy florist's auriculas were brought to Britain by Flemish weavers who settled in close-knit communities in Lancashire,

Auriculas have smooth fleshy leaves and clusters of velvety-looking zoned blooms on tall sturdy stems.

Yorkshire and Paisley in Scotland. Fleeing from religious persecution in their homeland, they not only brought their possessions with them, but their traditions too. These included the cultivation of florists' flowers, especially the auricula.

Many amateur hybridists worked with auriculas over the years, continually improving their form and constitution. It was recorded in 1757 that a green-edged auricula had appeared and another with a white edge. The development of the modern auricula was under way, pushed along by Lancashire silk weavers who have been given credit for the production of the show auricula in its present form, a genetically complex achievement. Despite fluctuations in their fortunes auriculas have always attracted a loyal following. They are likely to go from strength to strength with the benefits of modern micropropagation techniques.

SPECIES FROM EUROPE AND ASIA

The same could be said of the European primulas. The popularity of rock gardens and alpine plants has pushed primulas to the fore. They are such attractive and easily grown plants whose needs can easily be met by most home gardeners. Only since the turn of the century has much attention been paid to them individually. They were previously regarded as just material for breeding. This has changed rapidly with the general gardening public in the last 30 years or so, and species or primary hybrids grown for their own beauty can now be seen in a great many reasonable-sized gardens.

But the Asiatic primulas – exotic beauties in a wide range of forms, sizes and colours – have caused the biggest stir in gardens. They are now probably the widest cultivated of all primulas. Yet they have only been introduced during the past 150 years, often accompanied by lyrical stories. One of the most interesting concerns *P. sonchifolia*, a beautiful plant whose seeds remain viable for only a very short time and do not respond to ordinary storage. Discovered by Père Delavay in Yunnan in 1884, it was not cultivated in Europe until 1924, when some seeds were suc-

cessfully transported in a vacuum flask.

The same plant figured in a dangerous expedition by a forest ranger called Sukve in north Burma in 1930. Dormant plants were removed from ice-bound ground and packed among bamboo stems filled with ice, then despatched in the refrigerated hold of a steamer. The plants were presented to King George V and in the early spring of 1931 were in flower at the Royal Horticultural Society. A story of great care and devotion typical of many that enable us to enjoy these gorgeous exotics.

However, Mr. Bulley, founder of Bees' nursery, now the site of Liverpool

One of the most popular alpine-house primulas is Primula allionii *'Crowsley Variety'. The flowers are noted for their conspicuous white 'eyes'.*

University's Ness Botanical Garden, was probably responsible for introducing more Asiatic primula species than anyone else. He not only grew large numbers of these plants to perfection, but sent plant hunters out into the wild to bring back new discoveries. He and his nursery are commemorated in the names of the lovely *P. bulleyana* and *P. beesiana* – a tribute deserved by a great primula grower.

TENDER PRIMULAS

Diseases are not such a problem with the tender primulas, even though they have been cultivated for many years. The first indoor primula to arrive in Europe was *P. sinensis*. This was brought from gardens in Canton, China during the 1820s where it is believed to have been in cultivation for many centuries. Indeed it has been grown by man for so long that its origins are lost in the mists of time. Sixty years after *P. sinensis*, the lovely *P. obconica* made its debut. A native of China, it grows in the wild from the western province to the Sikkim Himalyas and southwards to the Shen States. Of all the indoor primulas, this has probably been the most developed by man, the large brightly coloured blossom of modern hybrids bearing little resemblance to the paltry pale lilac or purplish flowers of the species. Indeed today the plant breeder's skill has provided the gardener with cultivars which include frilled and bi-coloured blossoms as well as strong self colours.

Similar work has been done with *P. malacoides*, a little gem from Yunnan which made its first appearance in 1908. It was originally introduced in its single form, but by 1914 large-flowered double varieties were recorded. Today, it is available in myriad forms and is one of the finest primulas for the modern home. *Primula × kewensis*, although thought to be of recent introduction, owing to the new found enthusiasm of seedsmen, predates *P. malacoides* by ten years. A hybrid between *P. flori-bunda* and *P. verticillata*, it first appeared at the Royal Botanic Gardens, Kew, in a batch of seedlings. The first cross was sterile, but the *P. × kewensis* of today is fertile and now widely grown.

Looking at the array of tender primulas currently available, it seems unlikely that much improvement can be made, especially given the general lack of new and interesting half-hardy species for the plant breeder to work with. Doubtless, the colour range could be improved and maybe stature altered, but unlike their hardy cousins, the indoor primulas have probably reached their zenith.

RECENT DEVELOPMENTS

What of the present? Polyanthus and primroses are continuing to develop rapidly. The oil crisis of the 1970s led many commercial glasshouse growers to review their cropping schedules and turn to plants with a low temperature régime and small energy input. Polyanthus, and especially primroses, were targeted as suitable candidates. Breeding programmes already well advanced were further developed so that we now have superb varieties with neat compact foliage and tight flower heads in brilliant colours. In fact the plant breeder has turned our native primrose into a much loved houseplant, which now looks ill-at-ease in the open garden, visually and culturally. When contemplating coloured primroses for the garden, do be careful to avoid those brightly coloured strains that have been developed specially for the house plant trade.

Polyanthus have also been affected by

the move to low energy crops, though less radically. They are still thought of as garden plants, though many of the modern strains, like 'Cowichan', bear little resemblance to their forebears. Old cultivars are seeing a revival and modern micro-propagation techniques are doing the same for these parts of our heritage as they are for the auriculas of the Low Countries. It is not new developments in hybridizing that are important for the primula lover, but the means now available to rescue old forms and cultivars and market them at a reasonable price. Together with the ability to clean up virus-ridden stocks, this should ensure primulas remain one of our most popular garden plants.

CHAPTER TWO

PLANTING IDEAS

Most primulas associate happily with other plants that enjoy similar growing conditions. They also respond favourably in simulated wild conditions. The best example of this is the common primrose, *P. vulgaris*, a native of grassy banks and dappled shade. In a formal garden setting primroses look stark and ill-at-ease, but they are at home in a grassy sward, twinkling and smiling a welcome to spring.

While they are fairly tough, primroses do not enjoy fighting against coarse grasses, especially those like the thick mat-forming couch grass. It is not advisable to attempt planting young primroses through a grassy sward, unless it contains a preponderance of fine-leafed grass species.

To get primroses established successfully you need a well prepared site and a
bank clothed in grass raised from seed.
Planting primroses into a mixture of
unknown native grasses is imprudent,
for they are likely to be coarse, vulgar
and, like couch grass, spread freely with
strangling stolons. A mixture of fescues
and even a little rye grass will create a
sward in which primroses can grow
happily. They need planting of course,
for such a sward would not readily
accept seed. Pot-grown plants are the
easiest to get going, for the rootball
itself gives instant root territory. Divisions can be planted immediately after
flowering, but these need careful watering for some weeks afterwards to establish successfully.

BULBS FOR CONTRAST

Primroses in grass are a picture on their
own, but their beauty can be enhanced
by associating them with the white wood
anemone, *Anemone nemorosa*. The
white form of *A. blanda* is also charming, but though easier to establish than
its wild cousin, it lacks the primitive
charm of that species. Such anemones
flower in unison with primroses, though
in a cool spring primroses may extend
their flowering into summer. Make provision for this by planting bluebells
(*Endymion non-scriptus*) among them.
The fading lemon-yellow of the primroses and the azure shimmer of the
bluebells creates a lovely picture.

Cowslips (*P. veris*) and oxlips (*P. elatior*) are equally charming, but look
better on flat open swards. Mix them
with snakeshead fritillaries (*Fritillaria
meleagris*) and field pansy (*Viola arven-
sis*). Nothing looks finer in a gentle
spring breeze. Unlike primroses, which
do not naturalize freely from seed in
grass, the oxlip and cowslip multiply
quickly. But they are best started as
plants, ideally pot-grown.

FORMAL PLANTING

Polyanthus are derived from the previous characters, but do not respond to
the same kind of treatment. They are
essentially plants of the manicured garden and most readily accommodated in
a formal setting. Mixed colours are fine
for a cottage garden effect, but for
formal bedding arrangements the excellent modern strains in individual colours
are unsurpassable. A carpet of yellow
polyanthus, interplanted with sentries of
bright pink Darwin tulips, or a rich
velvety red with white Triumph tulips
can be highlights of formality in spring.

Achieving such a spectacle is not
difficult, provided good quality plants
and bulbs are used. Accuracy is para-
mount, every polyanthus being sited
so that by spring the foliage closes
together, yet the tulips can be planted
between. Some gardeners plant all the
polyanthus out first, then put the tulips
between, but this creates problems if
you cannot reach all parts of the bed
from the edge. It is difficult to pick your
way through the plants and get the bulbs
in accurately. Planting the bulbs with
the polyanthus as you progress across
the bed is the simplest option. To ensure
proper spacing, mark out the bed with

strings, and plant the bulbs with a planter that has accurate depths marked upon it. Getting the bulbs in at an even depth guarantees that marvellous spring spectacle of tulip blossoms all at the same height.

INFORMAL PLANTING

As already suggested, polyanthus are also good cottage garden plants, plants for mixed borders and tangled informality. Use the smaller flowered strains for the best effect. Large blossoming sorts like the Pacific Giants are big and blowsy and do not integrate well with such old favourites as yellow doronicum and double white arabis. The more modest polyanthus serve much like herbaceous plants and can easily be assimilated into a garden routine. There

is no need to lift and divide them every year – perfectly satisfactory results come from a three-year rotation.

GOOD CONTAINER PLANTS

The large-flowered and smaller blooming polyanthus can both be used in windowboxes and containers. Indeed, they are among the finest of spring-flowering plants for such situations. Provided they have a constant supply of moisture they will be happy. Drying out, even for a short period, is the greatest disaster likely to befall them. A single dry period ruins the plants for an entire season. Being resilient, polyanthus will seldom die completely, but it takes a year before the effects finally disappear.

Use well-grown pot plants, as their

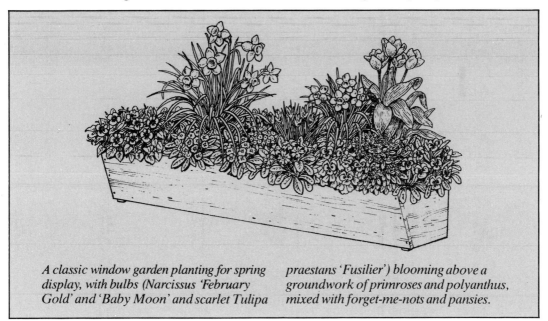

A classic window garden planting for spring display, with bulbs (Narcissus 'February Gold' and 'Baby Moon' and scarlet Tulipa praestans 'Fusilier') blooming above a groundwork of primroses and polyanthus, mixed with forget-me-nots and pansies.

A charming group of candelabra primulas set off by the fresh green fronds of a fern in a tub, suitable for early summer in shade.

root balls act as buffers against rapid desiccation. Windowboxes are best with small-flowered mixed polyanthus, the shorter the better, the higher up the building the windowbox is placed. Polyanthus on their own have much to commend them but they associate well with delicate spring bulbs, especially some of the later small cupped narcissi. The choice is yours, provided you adjust the planting density so all their roots can be accommodated.

Opposite: *A typical gold-centred alpine auricula, excellent for growing on the rock garden or in a border.*

Patio planters and containers are usually less exposed and so can take the less sturdy large-flowered strains. Single colour plantings are my favourites, but if the garden is informal a mixed array could be planted to good effect, especially in a cottage garden setting with a tub by the door. But neat white containers of formal design brimming with individual colours are more appropriate around modern homes. There is little opportunity for plant combinations in tubs, unless you have some existing structure planting. Polyanthus look good with evergreens, and conifers in tubs can be much enhanced by surrounding them with their bright colours. They co-exist happily, though the conifers benefit from top dressing and feeding when the polyanthus are removed after flowering.

Some alpine primulas like *P. marginata* are useful for growing in containers – not wooden barrels or plastic planters, but old stone sinks and troughs. Indeed, a miniature landscape could be contrived there. A gritty soil mixture and plenty of drainage ensure healthy growth. An occasional large stone adds to the landscape, which can be further enhanced by a dwarf conifer, like the pencil-shaped *Juniperus communis* 'Compressa', and colourful neighbours such as yellow-flowered *Saxifraga aizoon* 'Lutea' and the bristly green mounds of *Draba aizoides*: with careful thought, a very few plants such as these would create a delightful garden in miniature. The opportunities offered by a sink or trough are limitless and primulas can provide the colourful highlights.

OTHER USES

□ ROCK GARDENS

They provide reliable colour in the rock garden too. Do not plant them close together, but distribute them around the outcrops, planting several plants of each variety in a single pocket. Like mossy saxifrages and alpine phlox they are permanent and dependable. But their stiff formality makes them somewhat awkward to associate with other alpines. They go better with rock than contrasting foliage. Primulas such as 'Mrs J. H. Wilson' and 'Rufus' are superb against a backdrop of Westmorland limestone or millstone grit.

Of course many of the alpine varieties are excellent for scree beds and cultivation in the unheated greenhouse. The modern greenhouse often remains cold throughout the winter because of the excessive cost of heating. Provided there is plenty of ventilation, some of the more dainty alpine species like *P. farinosa* can be easily cultivated.

These respond well to pan culture in a free-draining gritty compost. Plant in groups for the best effect and arrange the pans neatly on the staging. While most gardeners display alpines under glass in serried ranks of pots and pans, with careful preparation an indoor landscape can be provided.

The most important factor in any such feature is the strength of the staging. Ordinary greenhouse staging is unlikely to be strong enough to take the weight of the quantity of rock and gravel necessary to create a pleasing miniature landscape. A solid benching with extra metal supports beneath is ideal, for then the gravel can be swept amongst strategically placed rocks in a natural fashion.

Apart from displaying the primulas themselves, it is important to build up a semi-permanent framework of background plants. Obviously individual varieties of primulas and other alpines have a limited flowering period, one variety following on from another. While these are regularly changed to ensure continuity of display, actually they are most effectively arranged as an integral part of a more permanent framework.

This background can consist of a wide range of dwarf shrubby plants, but some are much better than others. Few conifers associate happily, only one or two of the dwarf junipers and the smaller pines like *Pinus mugo* and its cultivars, although even this is a short term companion.

I like the emerging delicate foliage of the green and gold forms of the Japanese maples. *Acer palmatum* 'Dissectum' is superb, so too the golden *A. japonicum* 'Aureum'. These are lovely with some of the brighter coloured primulas, and because their foliage is not fully expanded at that time, the trees have a light, airy feel about them.

Trees of this sort have to be accommodated in large pots and do eventually outgrow their space. However, with a little care and regular repotting they can last for many years. The unheated greenhouse display should be regarded purely as a spring venture, the potted plants that comprise it being plunged outside during the summer.

☐ PEAT AND WOODLAND GARDENS

They suffer the same problems in the peat garden. So vivid and startling are their colours and so architectural their habit that they are best grown alone. Carpet the ground with creeping *Gentiana sino-ornata* if you like, but otherwise allow them to stand as colourful highlights. The use of a scrambling plant like the gentian disguises the fading foliage in late summer when the blue trumpets of this floral gem are just emerging. *Primula cockburniana* with orange-red blossoms in spring with the gorgeous blue *Gentiana sino-ornata* 'Angel Wings' in late summer provide dual garden highlights difficult to surpass.

Candelabra primulas are easier to work with, for most have a hint of the Orient and can be worked in with similar characters like hosta and ligularia. The deep red blossoms of *P. pulverulenta* on strong mealy stems are a dream when grown beside the cream and green variegated *Hosta undulata* 'Medio-variegata'. Both plants peak at exactly the same time, the hosta's variegated leaves being at their zenith the moment the primula opens its blossoms. No matter how early or late the season, these two plants come out in unison.

A variation of *P. pulverulenta* known as 'Bartley Strain' is equally lovely, a soft pink kind perfect for lightening a damp shady corner. This needs planting in a sweeping mass for the best effect, single plants looking naked in their loneliness. Ferns associate well with 'Bartley Strain' too, particularly the fine filigree forms of the lady fern, *Athyrium filix-femina*. These are a soft pea-green, gracefully arched and enjoy precisely the same conditions as candelabra primulas.

This is why I always grow *P.* 'Inverewe' among these woodland beauties. The finest of all candelabra primulas, this is too brightly coloured for its own good – bright orange-red with conspicuously mealy stems. A wonderful plant, but its colour is more suitable for bedding salvias. It will not mix happily with other primulas, for none can complement its vivid hue. Only among ferns and in the company of blue meconopsis does it look at all comfortable. Lady ferns are the perfect companions, but a carefully contrived arrangement with the sensitive fern, *Onoclea sensibilis*, also works well. This tends to creep about a bit and needs keeping under control, but its apple-green fronds with rich autumn tints provide an excellent foil for this moisture-loving star.

☐ BOG AND WATERSIDE

Few gardeners can afford to plant bold splashes of waterside or bog garden colour. Usually the area devoted to these plants is relatively small. To suit everyone's taste a mixed strain could be grown. The most popular, the 'Harlow Car' Hybrids, has something for everyone, with colours ranging from pale pinks and purples to red, orange and yellow. But such a kaleidoscope of colour needs careful placing, so a green background of shrubs or conifers and a

Above: *Border auriculas are reliable, free-flowering plants for garden decoration and often have attractive mealy foliage. Grow them in a border or on a rock garden.*

Opposite: Primula marginata *is an easy to grow, spring-flowering hardy perennial for the rock garden or alpine house. The leaves are coated with white meal.*

strong association with delicate looking ferns is necessary if the arrangement is to rest easily on the eye.

Where more room is available single-colour groups of deep orange *P. bulleyana*, bright red *P. japonica* and cool icy *P.j.* 'Postford White' can be established. These rarely blend happily when set next to each other, so make bold plantings of hostas like *H.sieboldii, H.lancifolia* and *H.*'Thomas Hogg' to contrast with and complement them. These all have fine leaves which reach perfection as the primulas flower, then add their own touch of colour afterwards with soft lilac or white bells on delicate flower wands.

EARLY AND LATE FLOWERERS

Moisture-loving primulas are not all late spring and early summer flowering plants. There are species and varieties for each end of the season. The tiny *P. rosea* and its vivid cultivar 'Delight' often flower during early spring through a coating of frost. Of diminutive stature, these little beauties are the first sign of life at the poolside. After flowering they can be lost in the foliage of marsh marigolds, bog beans and other water-side plants.

The lilac-flowered drumstick primula, *P. denticulata*, and its variously coloured cultivars similarly brighten any damp corner during early spring. Come snow or hail they stand their ground, bold sentinels against the vagaries of our weather. In contrast to *P. rosea*, the drumstick primulas leave a legacy of ugly cabbage-like leaves after flowering. So try to plant something bold and interesting in front of them to disguise their hideousness. A stout group of *Rodgersia aesculifolia*, with handsome chestnut-like leaves and frothy white flowers answers this problem. This not only extends the period of interest in this part of the garden, but is nowhere to be seen when the primulas are in flower.

The later part of the season can be covered by *P. sikkimensis* and *P. florindae*. The latter in particular will produce flowers until the first sharp autumn frosts. Both are like large cowslips, the Himalayan *P. florindae* growing up to a metre tall. They make marvellous pool or streamside plants with bold cabbage-like foliage and pendant sulphur blossoms on occasions colonizing the mud to the water's edge. Of somewhat boisterous disposition, they hold their own well among the earlier flowering bog plants, which become very leafy and unwieldy towards the close of summer. Both benefit from a green background, their iridescent blossoms not associating easily with astilbes, filipendulas and ligularias which flower about the same time. A bold stand of *Peltiphyllum peltatum* makes a lovely foil, its rich green leaves taking on a hint of autumnal bronze.

Primulas are wonderful garden plants, but because of their outstanding colours and startling architectural qualities are often grown alone. They are highlights in the garden. Whether at the stream or poolside, in the rock garden or in the grass, associated with tulips or alongside hostas, they are the focus of attention. When planting primulas look for beautiful foils rather than happy associations. For me they are the queens of the garden.

INDOOR ARRANGEMENTS

In the home or office, the tender species and their cultivars make a brave show. Again they do not happily associate with other pot plants as they are so bright and colourful. Only with foliage plants like the feathery indoor asparagus and the bold green ladder ferns are they in harmony. Grow them alone in solitary splendour, especially the large _P. obconica_, a houseplant that can stand on its own on any size windowsill. Be careful that it is not in a position where it can be accidentally brushed against, for many people suffer from an irritating rash if their skin comes into contact with the leaves. _Primula malacoides_ presents no such problems and, unlike the majority of other indooor primulas, can be used selectively in mixed arrangements. Associated with forced daffodils they present a lovely picture, only to be equalled by the common pink-flowered species when providing a frothy blush background to dark blue cinerarias. Even though visually difficult to mix with other indoor plants, primulas are amongst the most important, their principal virtue being their ease of cultivation and adaptation to any situation.

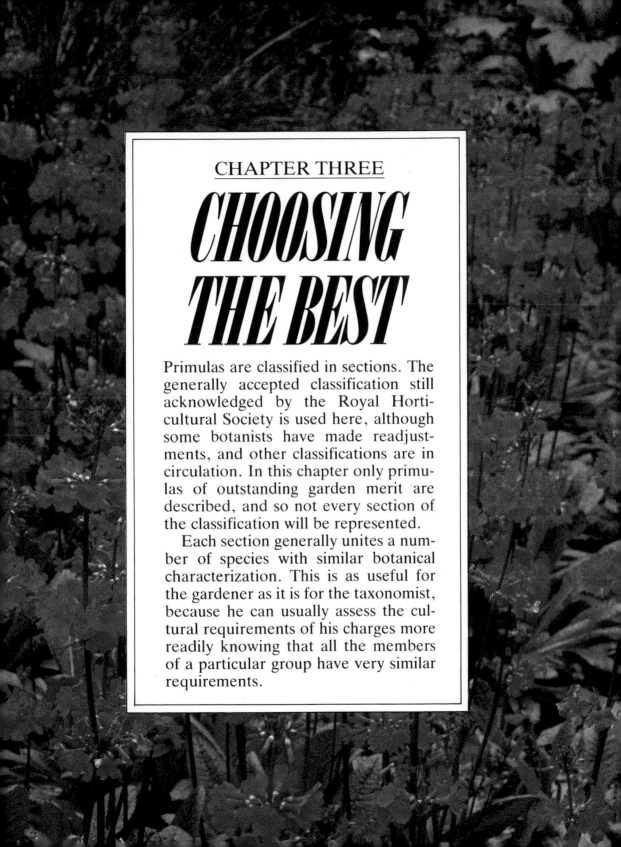

CHAPTER THREE

CHOOSING THE BEST

Primulas are classified in sections. The generally accepted classification still acknowledged by the Royal Horticultural Society is used here, although some botanists have made readjustments, and other classifications are in circulation. In this chapter only primulas of outstanding garden merit are described, and so not every section of the classification will be represented.

Each section generally unites a number of species with similar botanical characterization. This is as useful for the gardener as it is for the taxonomist, because he can usually assess the cultural requirements of his charges more readily knowing that all the members of a particular group have very similar requirements.

PRIMULA CLASSIFICATION

1 Amethystina	2 Auricula	3 Bullatae	4 Candelabra
5 Capitatae	6 Carolinella	7 Cortusoides	8 Cuneifolia
9 Denticulata	10 Dryadifolia	11 Farinosae	12 Floribundae
13 Grandis	14 Malacoides	15 Malvaceae	16 Minutissimae
17 Muscarioides	18 Nivales	19 Obconica	20 Parryi
21 Petiolares	22 Pinnatae	23 Pycnoloba	24 Reinii
25 Rotundifolia	26 Sikkimensis	27 Sinenses	
28 Soldanelloideae	29 Souliei	30 Vernales	

2.AURICULA

P. allionii

Spring-flowering hardy perennial primula for the alpine house or scree. Tubular flowers in rose, mauve or white, cover tight cushions of small bright green foliage. Height 10cm (4in), spread 10–15cm (4–6in). This is usually grown in an alpine house or frame where conditions can be much better controlled. It is vital to have a free-draining compost not too rich in nutrients. John Innes Potting Compost No. 1 with up to a third of sharp grit added is ideal, but plenty of broken crocks or gravel should be placed in the bottom of the pan. When planting allow enough room around and beneath the plant to spread a thin layer of fine grit. This helps to prevent the plant rotting off during the winter.

Careful watering is essential, water being dispensed around the edge of the pot or from beneath. Do not allow water to collect on the foliage. The compost must be kept moist, but not wet and ventilation must always be given to avoid botrytis. This is not common on primulas, but does affect *P. allionii*, decomposing the foliage and developing unsightly fungal growth. As it is vulnerable to fungal attack, all faded blossoms should be removed immediately. Grow tight in the pan and repot every year.

Primula allionii can only be grown successfully outdoors in a scree or the crevice of a rock outcrop. It is certainly not a plant for any ordinary pocket on the rock garden. During winter protect plants from the wet with a small sheet of glass raised on two bricks. Under glass or in the open, the plants benefit from light shading during very hot weather.

Propagation from seed results in very variable plants, though it is one of the quickest and easiest ways of raising a quantity of plants. Increase named cultivars from divisions removed immediately after flowering and repotted individually, or short stem cuttings which will root readily in a very sandy compost if taken during the summer. Though they need a degree of care and attention to detail, *P. allionii* and its myriad named varieties are among the most popular alpine primulas.

P.a. alba The name given to a range of variable white forms. In every other respect similar to the species. Height 10cm (4in), spread 10–15cm (4–6in). The following cultivars derived directly from *P. allionii* have the principal characteristics, including height and spread, of the species.

P.a. 'Anna Griffith' A neat compact form discovered in the wild with bright green toothed leaves from among which emerge very pale pink blossoms with notched petals.

P.a. 'Apple Blossom' Large delicate rose-pink blossoms shading to white. Distinctive wavy edged petals. Broad bright green leaves with toothed margins.

P.a. 'Avalanche' A selection from *P.a. alba* with beautiful pure white rounded blossoms. Green leaves with toothed margins.

P.a. 'Crowsley Variety' One of the most popular alpine primulas. Flowers deep crimson with a conspicuous white eye. Leaves small, grey-green and sparingly toothed.

P.a. 'Mary Berry' An unusual squat mat-forming, rather than mound-forming variety. Flowers large, dark reddish-purple. Foliage pale green and slightly toothed.

P.a. 'Pinkie' One of the earliest flowering widely grown cultivars. Forms a tiny neat cushion smothered with lilac-pink blossoms.

P.a. 'Praecox' The earliest flowering *P. allionii* cultivar. Often starts to bloom before the turn of the year. Not of the best conformation, but a welcome early addition to the alpine house, sporting bold lilac-pink flowers.

P.a. 'Snowflake' A large-flowered white cultivar occasionally flushed with pink. Distinctive overlapping petals with notched tips and wavy edges.

P.a. 'Vicountess Byng' Purplish-pink blossoms with a white eye borne among handsome grey-green foliage. Not freely available commercially now, but circulating among enthusiasts and well worth looking out for.

P.a. 'William Earle' Very large lilac-pink flowers with a clear white eye. Very broad wavy-edged petals.

□ *P. allionii* HYBRIDS
Primula allionii is a promiscuous plant, hybridizing with at least seven other species and yielding an array of very interesting alpine plants. Several of them show a trace of the distinctive farina inherited from the other parent.

P. 'Beatrice Wooster' (*P. allionii* × *P.* 'Linda Pope') Very similar to *P. allionii* in stature and habit, but slightly farinose flower parts and blossoms of clear pink with a white eye. Leaves mid-green.

P. 'Ethel Barker' (*P. allionii* × *P. hirsuta*) Bright carmine blossoms with a white eye borne in groups of three or five. Leaves green and downy.

P. 'Fairy Rose' (*P. allionii* × *P.* 'Linda Pope') Large rose-pink blossoms above conspicuously toothed green leaves.

P. 'Gladaline' (*P. allionii* ×
*P. pubescens***)** Groups of cherry-red
blossoms with bright yellow eyes.
Leaves shiny green, toothed and with a
dusting of farina, produced in neat, tight
rosettes.

P. 'Ivanel' (*P. allionii* ×
*P. carniolica***)** This is a selected reddish-
purple form of the cross which has been
given a name and is now propagated
vegetatively.

Above: Primula marginata *'Beatrice
Lascaris' is a slow-growing form, ideal for
the rock garden or alpine house.*

Opposite: Primula *'Linda Pope' is a*
P. marginata *hybrid and one of the best-
loved of the alpine primulas.*

P. 'Joan Hughes' (*P. allionii* × *P.* **'Linda
Pope')** Deep magenta blossoms darken-
ing towards the edges with a small white
eye. Leaves produced in neat cushions.

P. **'Lismore Yellow'** (*P. allionii* ×
P. auricula) Of similar habit, but more
compact than *P. auricula*. Soft yellow
blossoms.

P. **'Purple Emperor'** (*P. allionii* ×
P. **'Linda Pope'**) A neat, low growing,
compact plant with scalloped green
leaves and rich purple blossoms with
tiny white eyes.

P. auricula
(Auricula) Spring-flowering perennial
alpine plant with fragrant yellow flowers
borne in a similar arrangement to
garden polyanthus. Leaves green, soft
and smooth, often liberally coated with
whitish or pale yellow farina. Height
15–20cm (6–8in), spread 15cm (6in).
An easily grown alpine plant for the
sunny rock garden pocket which is
free-draining, yet offers a constant
supply of moisture. In drier conditions
they benefit from partial shade. It is
possible to raise *P. auricula* from seed,
but the progeny are likely to be variable. Mixed hybrids are often offered by
seed companies. Good forms can be
increased by division immediately after
flowering. Pieces of plant with long
stems can be taken as cuttings in summer and root well in a very sandy
medium.

In the gardener's hands the auricula
has over the past few centuries been
transformed into a quite remarkable
flower. Indeed auriculas are now classified in very distinct divisions. These are
particularly important to exhibitors who
form the majority of enthusiastic auricula growers. As with chrysanthemums/
dahlias the characteristics are so diverse
that each class has its own standard.

Show auriculas These have brightly
coloured blossoms with a distinctive
circle of meal known as paste
surrounding the tubular part in the
centre of the flower. There are four
major groupings: Green-edge; Grey or
White-edge; Fancy; and Show Selfs.

Green-edge
These are characterized by petals with a
black ground colour, but an outer edge
of green. Leaves are not mealy.

'Chloe' Said by enthusiasts to be the
best green of all. Six or seven large
blossoms to a truss. Broad, rounded
bright green leaves with serrated
margins.

'Roberto' A very free-flowering green,
rather smaller than its contemporaries,
but producing myriad blossoms.
Exhibitors do not recommend this for
beginners to the show bench, but it is
not a difficult plant to grow for pure
pleasure.

Grey or White-edge
A group whose petals are characterized
by a black ground colour, the outer edge
being dusted with meal of varying density. Leaves usually mealy, sometimes
green.

'Helena' A wonderful grey with rich
yellow tubes. Leaves heavily coated
with meal and having distinctly serrated
edges.

'The Bride' Intense black ground

colour with beautiful white edge. Leaves are only serrated at the tips and liberally covered with meal.

Fancy
Flowers with a ground colour other than black. Cultivars in green, grey or white-edged forms. These are generally regarded by enthusiastic exhibitors as misfits, but they do have a charm of their own and are perfectly adequate plants for garden decoration.

'Coffee'   An attractive yellow with a bold grey edge.

'Rajah'   Scarlet blossoms with distinctive green edge.

Show Selfs
These have blossoms of a single self-colour and variable mealed foliage.

'Blue Nile'   One of the best blues, although some experts say that a real true blue does not exist. Very fine white paste. Foliage mealed.

'Cherry'   As its name suggests, a cherry-red flowered cultivar. Foliage mealy.

'Everest Blue'   A good strong violet-blue flowered plant. Large blossoms, mealy foliage.

'Guinea'   A very reliable yellow-flowered cultivar with very heavily mealed foliage.

'Pat'   A somewhat temperamental beauty which requires careful watering. Gorgeous red blossoms and handsome mealy foliage.

'Sheila'   A vigorous yellow-flowered cultivar of good conformation. Leaves evenly covered with meal and slightly serrated.

'The Mikado'   One of the oldest cultivars still in general cultivation. Dark purplish-black blossoms with a yellow centre. Leaves yellowish-green and without meal.

Alpine auriculas   These are considered by enthusiasts an inferior group of auriculas, but for the ordinary home gardener they are a delight. Of easy cultivation and not such stiff formal appearance as their show counterparts. Alpine auriculas are divided into gold-centred and light-centred cultivars.

Gold-centred

'Bookham Firefly'   A well established cultivar with blossoms shading from crimson to maroon with bright golden-yellow centres. An excellent plant for the border or rock garden.

'Winifred'   Red blossoms with golden centres. Often used for show work, but quite amenable to garden cultivation.

Light-centred

'Argus'   One of the oldest auriculas still in cultivation. Originating last century, it has held its own in gardens and on the show bench. Flowers plum-purple with white centres.

'C. W. Needham'   Dark purple-blue with a white centre. A well-tried and reliable variety.

Opposite: *An excellent companion for* Primula beesiana *at the waterside is* P. bulleyana, *a candelabra species.*

Above: Primula × pubescens *is easily grown, given sun, good drainage and ample moisture. The brilliant 'Rufus' is shown here.*

The cultivation of auriculas outlined earlier is insufficiently precise for the production of quality plants for the show bench. It is not my intention here to move into the realms of the exhibitor, but by following some of his techniques higher quality plants can be obtained for general decoration. The cultural regime previously described covers the general cultivation of auriculas as rock garden or border subjects.

Most show and alpine auriculas produce the best results when grown in pots and most growers still prefer clay pots and soil-based compost, rather than modern plastic pots and soilless peat composts. John Innes No. 2 potting compost with up to 25% sharp grit added yields most satisfactory results, although keen showmen have their own formulae including cow dung and seaweed. The plants must be repotted regularly, preferably directly after flowering. They can spend the summer and early autumn outside in a plunge, but take care to see that they do not get too wet. Move them to the greenhouse or sun lounge in autumn and keep as cool as possible throughout the winter. Flowering will start in early spring. During the winter take great care with watering and keep a close watch for fungal diseases. Regular removal of fading leaves and blossoms will help prevent the incidence of disease. Whenever possible give maximum ventilation. Auriculas do not respond to forcing and should be kept as cool as possible at all times.

Border auriculas For gardeners like me who are not ardent exhibitors, these are the favourites. Reliable free-flowering plants for garden decoration, equally at home in a border or the rock garden.

'Broadwell Gold' Gorgeous frilled golden-yellow blooms with a paste centre. Foliage liberally coated with meal. A first-class cultivar for spring bedding schemes.

'Linnet' An interesting late-flowered cultivar with dull greenish-brown and ochre blooms. More of a curio than a plant of beauty, although it is much-loved by many growers.

'McWatts Blue' A fragrant strong blue-flowered cultivar with very mealy foliage.

'Old Irish Blue' An extremely popular velvety-blue flowered plant with mealy foliage. There are a number of plants around under this name, however, which appear to be poor quality seedlings.

'Old Red Dusty Miller' Large scarlet blossoms with a white eye on a yellow background. Mealy foliage. Ideal for windowboxes and containers.

'Old Yellow Dusty Miller' Bright yellow flowers with a conspicuous white farinose centre. Mealy foliage. A good subject for planters and windowboxes.

'Royal Velvet' Deep velvety red to maroon blossoms with a creamy-white centre. A reliable garden plant.

While border auriculas will respond to the conditions outlined for the ordinary species, they can be grown in a wider range of situations. As their name suggests they make first-class spring-flowering border plants. They are also excellent for tubs, planters and window-boxes. Most garden soils that are in good heart and contain a liberal quantity of organic matter will grow good auriculas. While tolerant of an open situation they will not live happily under a blistering sun, especially if nearby shrubs and plants are depleting the soil of moisture. Cultivation in partial shade generally results in the best blooms. Regular replanting is essential to maintain quality. Divide plants after flowering at least once every three years.

P. carniolica

A most attractive spring-flowering hardy perennial primula often cultivated in the alpine house in pans. Flowers fragrant, funnel-shaped, rose-pink to purplish-pink, occasionally white. Leaves broad, green, not unlike those of *P. auricula*, but without farina. Height 10–20cm (4–8in), spread 15–20cm (6–8in). An easily grown alpine primula in a well drained gritty compost in a pan. John Innes No. 1 potting compost with up to 25% by volume of grit added is ideal. Top-dress the pan with grit. Does not enjoy full sun, so a little shading is desirable, as is a constant supply of moisture. Not a reliable plant in the open rock garden, often turning yellow and dying back during the summer. Cultivated indoors it grows lustily and can easily be increased by division immediately after flowering. Seeds rarely set and are only occasionally offered in the seedlists of specialist societies.

P. clusiana

One of the most widely grown small alpine primulas. Spring flowering with large rose or lilac blooms on stout stems on small compact plants. Small, broad glaucous green foliage in neat rosettes. Not unlike a small form of *P. auricula*. Height 10cm (4in), spread 10cm (4in). A useful plant for pan culture in the alpine house where it will respond to John Innes No. 1 potting compost with up to 25% by volume sharp grit added. It should be able to hold its own in a suitable open position in a free draining soil on the rock garden. Like most other primulas in this section it does not enjoy a sun-baked spot. It also benefits from a generous layer of gravel or sharp grit scattered around the base of the plant to prevent crown rotting during the winter. Although this species can be raised easily from seed the resulting seedlings are likely to be extremely variable. Careful division of the plants immediately after flowering is the best method of propagation and also benefits the plant. Cuttings can be taken during the summer and rooted in very sandy compost.

P. glaucescens

A lovely little spring-flowering alpine primula with rose or pale pink tubular blossoms with distinctively notched petals. These are borne on short, stout flower stalks which arise from congested star-like rosettes of glabrous green lanceolate foliage. Height 5–15cm (2–6in), spread 10–15cm (4–6in). This can be a shy flowering plant so it is important to select a good clone when it is in bloom.

Opposite: *A charming candelabra for the peat garden,* P. cockburniana *blooms in early summer. It is short-lived, though, and should be regularly reproduced from seed.*
Above: *The famous 'Harlow Car' hybrids flower from early spring into summer, and embrace almost every colour.*

The best forms often produce a second flush of blooms each year. Grow in pans of free-draining gritty compost in an alpine house, or in a pocket of free-draining soil on the rock garden. It will grow in an open position, but should be shaded from long periods of hot sunshine. Division of the plants every two years or so maintains good vigour and provides a ready means of propagation.

P. marginata
Spring-flowering hardy perennial for the

rock garden or alpine house. Heads of funnel-shaped lilac-blue flowers above toothed green leaves often heavily overlaid with white meal. Height 10–15cm (4–6in), spread 10–20cm (4–8in). This is one of the easiest of the choice rock garden primulas to grow, provided drainage is adequate. It is happiest in a rock crevice or scree, but will exist contentedly in an open rock garden pocket. The main disadvantage of growing *P. marginata* in the open is that it loses much of the attractive farina on its foliage in our damp climate. Under glass, where it is protected from rain, this greatly enhances the beauty of the plant, especially while it is not flowering. In the alpine house grow it in pots or pans of free-draining compost. John Innes potting compost No. 1 with up to 25% by volume of sharp grit provides a perfectly adequate medium. When well established, clumps of *P. marginata* tend to become leggy and seem to push out of the soil. It is quite in order to lift, divide and replant them, burying the stem in the soil sufficiently to give the plant a pleasing appearance. Propagate by division immediately after flowering.

□ *P. marginata* VARIETIES AND HYBRIDS

P.m. alba As with many alpine primulas, the name *alba* refers to a collective of white forms rather than an individual variety. Most forms tend to turn pinkish with age.

P.m. 'Barbara Clough' Named after a great amateur gardener and benefactor, this has pinkish-lilac flowers with a white eye. Leaves broad, toothed and coated in yellowish farina.

P.m. 'Beatrice Lascaris' Blue-flowered cultivar with a white mealy eye. Leaves broad, green with small teeth and yellowish farina. A slow-growing form believed by some gardeners to be a hybrid including *P. allionii.*

P.m. 'Coerulea' Blue funnel-shaped blossoms. Leaves green, but heavily coated in white farina.

P.m. 'Elizabeth Fry' An old wellestablished cultivar with large silverylavender blooms borne in small groups on short flower stems. Small green, toothed leaves with yellow farina.

P.m. 'Holden Clough' Small blue funnel-shaped flowers on a compact plant. Leaves narrow, regularly toothed and heavily coated with white meal.

P.m. 'Kesselring's Variety' One of the later flowered varieties with deep lavender blossoms above mounds of green leaves with white farina.

P.m. 'Prichard's Variety' One of the loveliest varieties. Large lilac-purple blossoms with striking white eyes in a neat umbel on strong flower stems. Leaves green-edged with white farina.

P.m. 'Rubra' Medium-sized blossoms of deep rosy-lilac above broad green toothed leaves which are dusted with a little yellow farina.

The following are a group of hybrids which include contributions from *P. allionii, P. latifolia, P. auricula,*

P. pedemontana and *P. villosa.* A number are also crossed between *P. marginata* and some of its hybrids.

P. 'Hurstwood Susan' (*P. marginata* × *P.* 'Decora') An award-winning hybrid with umbels of scented violet-purple flowers with white eyes. Leaves green, somewhat rounded and lightly dusted with farina.

P. 'Linda Pope' (*P. marginata* hybrid) Probably the best loved of these alpine primulas. Large flowers of rich mauve-blue with white mealy eyes. Leaves large, toothed, green with a heavy dusting of white farina.

P. 'Marven' (*P. marginata* × *P.* × *venusta*) Another very popular hybrid which is widely cultivated and ideal for the newcomer to primulas. Deep violet-blue flowers with dark eyes surrounded by a small white halo. The leaves are light green and lightly dusted with farina, especially along the margins.

P. 'White Linda Pope' A white form resulting from a seedling of 'Linda Pope'. Pure white blossoms, pale green in bud, open to reveal a pale greenish-yellow eye. Large green rounded leaves.

P. minima
A challenging, but lovely hardy, spring-flowering alpine plant for the enthusiast sporting rose-pink blossoms with a white eye in small umbels. Compact rosettes of leathery green leaves with deeply serrated ends. Height 5–10cm (2–4in), spread 5–10cm (2–4in). Not difficult to cultivate, but often shy-flowering in cultivation. Can be grown in a pan in the alpine house, or outdoors in a very well drained pocket on the rock garden. A lime lover, it seems to benefit from a dusting of garden lime mixed into the soil or compost at planting time. A free-draining medium is essential, a mixture of John Innes potting compost No. 1 with up to 25% by volume of sharp grit added being a good foundation for pan culture. Tolerating full sun, it does not enjoy being sun-baked, and needs the provision of a little shade and a constant supply of moisture. There are a number of forms, probably hybrids of *P. minima*, with improved flowering qualities. These can be easily increased from division or cuttings. Availability and subsequent germination of seeds are erratic, even freshly sown seeds often needing a year before germination is complete.

P. × pubescens
An important group of hybrids widely cultivated as rock garden or alpine house plants, the result of unions between *P. auricula*, *P. hirsuta*, *P. villosa*, *P. rubra* and probably many others. They are all of similar height and habit and benefit from the same cultivation as *P. auricula*. Height 10–15cm (4–6in), spread 15–20cm (6–8in).

P. × p. alba A collective name given to myriad white forms. Some have been selected and named.

P. × p. 'Bewerley White' A strong-growing creamy-white flowered cultivar. Fresh green leaves.

P. × p. 'Boothman's Variety' One of the most popular and easiest to grow.

Neat, compact habit. Flowers bright crimson with a white eye. Petals conspicuously overlapping and distinctly notched.

***P. × p.* 'Faldonside'** An old favourite long in cultivation. A reliable cultivar with reddish-pink flowers with white eyes.

***P. × p.* 'Freedom'** A strong-growing deep lilac flowered cultivar with bold dark green, toothed leaves.

***P. × p.* 'Harlow Car'** A vigorous and easy award-winning plant developed at the gardens of the Northern Horticultural Society near Harrogate. Heads of large creamy-white flowers above lusty mid-green foliage.

***P. × p.* 'Mrs J. H. Wilson'** An old and very popular cultivar. Flowers fragrant, purple with a white centre and borne in many-flowered umbels above compact mounds of greyish-green leaves.

***P. × p.* 'Ruby'** A charming small-flowered hybrid with wine-red blooms with conspicuous white centres.

***P. × p.* 'Rufus'** Bright red blooms with golden centres are borne in many-flowered groups on strong flower stems. Foliage pale green, shallowly toothed and in neat rosettes.

P. × venusta (P. auricula × P. carniolica)
Spring-flowering hardy perennial primula with crimson, purple or brownish-red fragrant blooms above leaves with a variable white farina. Height 10–20 cm (4–8 in), spread 10–15 cm (4–6 in). A lovely little hybrid well suited to the rock garden or pan culture in an alpine house. Grow in a free-draining gritty medium. Propagate by careful division of established plants immediately after flowering.

P. wulfeniana
Early spring-flowering hardy perennial with tubular deep rose flowers with distinctively notched petals. Lance-shaped, somewhat leathery dark green leaves. Height 5–10 cm (2–4 in), spread 10–15 cm (4–6 in). A lovely mat-forming primula best suited to pan culture in an alpine house or careful cultivation on a scree. A lime-lover that will prosper in John Innes potting compost No. 1 with a dusting of garden lime added and up to 50% by volume sharp grit. Can be increased by careful division after flowering.

3.BULLATAE

P. forrestii
Late spring or early summer flowering hardy perennial with many-flowered umbels of yellow blooms, each with a distinctive orange eye. Leaves ovate, mid-green, rather coarse. Height 15–20 cm (6–8 in), spread 15–30 cm (6 in–1 ft). A good plant for the alpine house where the moisture level in the compost can be controlled. Requires a free-draining, constantly moist organic

Opposite: *One of the best-known and most easily grown of the candelabra primulas is* P. japonica, *a strong grower blooming in late spring and early summer. A moisture-loving species for sun or shade.*

Primula forrestii is an early summer flowering charmer with orange-eyed yellow blooms. It is best in an alpine house in constantly moist but free-draining compost.

stalk during summer. Leaves large, green, obovate and rather coarse. Height 30–60cm (1–2ft), spread 30–45cm (1–1½ft). Hardy perennial well suited to a moist position in sun or shade. Easily increased from seeds, especially if sown immediately after harvesting. Plants can be carefully divided during early spring or immediately after flowering.

P. aurantiaca

Early summer-flowered hardy perennial with bright reddish-orange blooms arranged in neat tiered whorls. Leaves green, long, broad and rather coarse, slightly aromatic. Height 60–90cm (2–3ft), spread 30–45cm (1–1½ft). An excellent and reliable plant for a bog garden or waterside. Revels in a richly organic damp medium. Increase from seeds, best sown immediately after harvesting. Packeted seeds sown the following year usually germinate erratically. Established plants can be divided as they emerge through the soil during early spring or immediately they finish flowering.

P. beesiana

A vigorous early summer flowering hardy perennial. Flowers rosy-carmine with a yellow eye borne in dense tiered whorls on stout flower stalks. Leaves large, green, cabbagy, up to 30cm (1ft) long. Height 60–75cm (2–2½ft), spread 45–60cm (1½–2ft). A wonderful strong growing plant for the bog garden or poolside. Enjoys a cool deep root-run in a damp richly organic soil. Grows well in sun or partial shade. Propagate by

medium, but not wet conditions – difficult to provide in the open. In the wild it tends to grow in much drier conditions than are successful under cultivation. Propagate from seeds sown as soon after gathering as possible, or by division of the spreading rhizomatous growths in summer, immediately after flowering.

4.CANDELABRA

P. anisodora

A strongly aromatic plant, in both flowers and foliage. Flowers brownish-purple with a green eye, borne in small tiered whorls around a strong flower division of the emerging crowns in spring, or from seeds sown immediately after harvesting. Packeted seeds sown during the spring usually germinate, but somewhat erratically.

P. bulleyana
Named after the great primula collector, this hardy perennial primula flowers during early summer. Large deep orange flowers arranged in tiered whorls on strong flower stalks. Leaves large, green and cabbagy. Height 60–90 cm (2–3 ft), spread 30–45 cm (1–1½ ft). An excellent companion for _P. beesiana_ and _P. burmanica_ at the waterside. Must have a cool damp root-run in a deep richly organic soil. Grows well in sun or partial shade. Can be increased by careful division in early spring just as the shoots emerge or immediately after flowering. Freshly gathered seeds germinate freely. Packeted seeds sown the following spring germinate erratically.

P. burmanica
Vigorous hardy perennial flowering during early summer. Large reddish-purple blossoms with conspicuous yellow eyes are borne in tiered whorls on strong flower stems up to 60 cm (2 ft) high. Leaves green, large and cabbagy. Height 60 cm (2 ft), spread 30–45 cm (1–1½ ft). Requires a good damp, deep, cool soil rich in organic matter. Will grow successfully in sun or shade. Propagate from careful division of the crown during early spring just as they are coming into growth, or immediately after flowering. Freshly gathered seeds germinate freely if sown immediately.

Older seeds usually germinate though erratically.

P. chungensis
A lovely hardy perennial primula with handsome tiered whorls of pale orange blooms during early summer. Leaves green, coarse, and rather cabbagy in large clumps. Height 60–75 cm (2–2½ ft), spread 30–45 cm (1–1½ ft). Revels in a moist richly organic soil in full sun or partial shade. Readily increased from seeds sown as soon as possible after harvesting. Older seeds will germinate, but somewhat erratically.

P. cockburniana
Charming short-lived hardy perennial with small vivid orange-red blooms on slender wiry flower stems during early summer. Leaves small, green, oblong-obovate in rosette-like clusters. Height 15–30 cm (6 in–1 ft), spread 10–15 cm (4–6 in). An excellent plant for the peat garden. Unlike many other species in this section, _P. cockburniana_ resents really wet conditions. The damp, humid environment of a well prepared and maintained peat garden suits them best. They are vigorous free-flowering plants, often appearing to flower themselves to death. Even removing fading flower stems has little effect in preventing their decline. This lovely primula should be regularly reproduced from seed and, as a matter of course, sown immediately it has ripened. Sometimes the plant will self-sow and perpetuate itself, but this is a hit or miss affair. Young seedlings of _P. cockburniana_ should always be pot

Above: *A great favourite with gardeners is the candelabra species* P. pulverulenta, *which flowers in spring and early summer. It grows in any moist soil and in sun or shade.*

Opposite: *A bog garden in early summer, made highly colourful with moisture-loving primulas. Acting as a foil for these are ferns and rushes.*

grown, for unlike most other primula species it does not like being disturbed when planted out. Likewise it is not very desirable to lift and divide this plant.

P. 'Harlow Car' HYBRIDS

This is probably the most famous strain of candelabra primulas ever developed. Flowering from early spring into summer they embrace almost every colour imaginable. A complex group of hybrids, they have large blooms, most with conspicuous yellow eyes, arranged in crowded tiered whorls on strong flower stems. The leaves are varying shades of green and rather coarse and cabbagy. Height 60–75 cm (2–2½ ft), spread 30–45 cm (1–1½ ft). This strain was developed over many years at the Northern Horticultural Society's famous gardens at Harlow Car, Harrogate. Their origins are unclear, but the strain is constantly being upgraded and improved by reselection. Plants for a wide range of situations, provided they have a cool moist root-run. The extensive plantings at Harlow Car embrace sun, partial shade and full shade, the hybrids doing tolerably well in all conditions. Propagate from seeds sown immediately after harvesting, or in spring. This yields a good strong mixture of colours. Especially fine colours that appear can be reproduced by careful division in early spring as the shoots start to appear or immediately after flowering.

P. helodoxa

Hardy summer-flowering perennial primula with the most vivid yellow open bell-shaped flowers in tiered whorls. Leaves bright green, long and tapering. Height 60–90 cm (2–3 ft), spread 30–45 cm (1–1½ ft). An excellent plant for a moist border or bog garden where the soil contains plenty of peat or well rotted garden compost. While enjoying dampness, it will not tolerate standing with its roots in water. Easily increased from freshly gathered seeds sown immediately. Packeted seeds retain little viability and relatively few seedlings are produced from their erratic germination. Division of the clumps in early spring just as they are coming into growth is satisfactory. So is lifting and dividing immediately after flowering.

P. 'Inverewe'

The most spectacular member of the *Candelabra* section and a must for all primula lovers. An early summer-flowering perennial with bright orange flowers in dense tiered whorls. The flower stems are heavily coated with white meal and contrast beautifully with the vivid blooms. Leaves plain green, rather coarse, more or less elliptical and rather cabbagy. Height 45–75 cm (1½–2½ ft), spread 30–45 cm (1–1½ ft). An excellent plant for a moist border or bog garden. Demands a good richly organic soil and constant moisture. It will grow happily in sun or partial shade. Should be divided in early spring as shoots are starting to push through the soil, or immediately flowering is over.

P. japonica

A strong-growing hardy perennial primula with bold flowers of deep red produced in tiered whorls on extremely stout flower stalks during late spring and early summer. Large cabbage-like leaves, coarse, rather smooth, green with a slight bluish tint. Height 45–75 cm (1½–2½ ft), spread 30–45 cm (1–1½ ft). One of the most reliable moisture-loving primulas. Demands a damp richly organic soil. Will grow in full sun or partial shade. Easily increased from seed, even packeted seeds producing tolerable results. Plants can also be divided during early spring, just as new shoots are emerging, or immediately flowering is over.

P.j. 'Miller's Crimson' Probably the finest form of *P. japonica*, the colour being particularly intense and uniform. In every other respect identical to the species. Height 45–75 cm (1½–2½ ft), spread 30–45 cm (1–1½ ft). Although a named cultivar, it can be raised from seed and will come true. Seeds sown immediately after harvesting are best, but packeted seeds also yield creditable results. Plants can be divided in early spring or immediately after flowering.

P.j. 'Postford White' This has flowers of a cool icy white with a conspicuous orange-yellow central ring. Sometimes the flowers show a hint of pink. A vigorous plant, the same in most respects as *P. japonica*, but with somewhat paler foliage. Height 45–75 cm (1½–2½ ft), spread 30–45 cm (1–1½ ft). Another cultivar that comes almost 100% true from seeds. Freshly gathered seeds produce the best results, but reasonable germination can be expected from packeted seeds. Plants can be lifted and divided during early spring or immediately after flowering.

P. poissonii
An interesting early summer flowering, hardy perennial primula with tiered whorls of deep purplish-crimson flowers with yellow eyes. Leaves oblong or obovate, coarse, glaucous green in neat clumps. Height 30–45 cm (1–1½ ft), spread 30 cm (1 ft). Requires a moist richly organic medium in sun or partial shade. Will not tolerate standing in water. Easily raised from seeds, especially if sown directly after harvesting. Plants can be divided in early spring or immediately after flowering.

P. pulverulenta
A great favourite with gardeners. Spring and early summer flowering with deep red blooms with conspicuous purple eyes. Bold tiered flower stems densely covered in white farina. Green cabbagy leaves. Height 60–90 cm (2–3 ft), spread 30–45 cm (1–1½ ft). One of the finest moisture-loving primulas. A hardy plant of great versatility and reliability growing in any moist, richly organic medium in sun or shade. Easily increased from seeds. Readily divisible in spring when its shoots are just peeping through the soil, or in midsummer after flowering is over.

P.p. 'Bartley Strain' A fine pink-flowered strain that is the same in every other respect as *P. pulverulenta*. Height 60–90 cm (2–3 ft), spread 30–45 cm (1–1½ ft). Easily raised from seed, although established plants can be divided in spring as evidence of growth is seen, or alternatively after flowering.

P. smithiana

An early summer-flowering hardy perennial species with handsome pale yellow cup-shaped flowers in dense whorls. Leaves long, narrow and bright green. Height 30–45 cm (1–1½ ft), spread 30 cm (1 ft). A pleasing plant for a moist corner in the garden. *P. smithiana* also makes a good background plant for the larger peat garden. It is not well adapted to bog garden conditions although it does enjoy a moist richly organic medium. Increase from division of the crowns in spring just as new shoots are appearing, or during the late summer when flowering has finished.

The edges of a stream make an ideal home for many of the moisture-loving primulas. The massive leaves in the background belong to lysichitum, another dramatic bog plant.

5. CAPITATAE

P. capitata

Mid to late summer-flowering hardy perennial with small deep violet bell-shaped flowers arranged in rounded flattened heads on short, stout flower stems. The small rosettes of lance-shaped green leaves are heavily white farinose, especially beneath, as are the flower stems. Height 15–30 cm (6 in–1 ft), spread 15–20 cm (6–8 in). A beautiful plant for the peat garden or a shaded pocket in richly organic soil on the rock garden. Does not divide easily, as individual plants tend to grow larger without producing progeny. Easily raised from seed, especially when freshly gathered and sown straight away.

P.c. mooreana Very similar to the species but flowering over a slightly later period. Flowers open in a more flattened head, leaves green above, densely white farinose beneath. This plant is often offered in alpine nurseries as _P. mooreana_. A more robust and easily grown form. Height 15–30cm (6 in–1 ft), spread 15–20 cm (6–8 in). This is another plant for the peat garden.

P.c. sphaerocephala Another variation of the species sometimes offered by alpine specialists. Funnel-shaped deep violet flowers in tight globose heads during mid and late summer. Leaves lance-shaped, green above and beneath. Height 15–30cm (6in–1 ft), spread 15–20cm (6–8in). A good peat garden plant.

7.CORTUSOIDES

P. cortusoides
Spring-flowering hardy, but often short-lived perennial. Flowers small, purplish-rose in many-flowered umbels. Leaves ovate-oblong, soft green and downy. Height 15–30cm (6in–1ft), spread 20–30cm (8in–1ft). A very easily grown plant for a shady spot. Ideal for a peat garden, but will grow tolerably well in a wild garden provided that there is not too much competition. Plants are lusty and healthy for two or three years, but then start to deteriorate and often die out. Replace regularly from seeds which germinate reasonably freely at any time.

P. heucherifolia
Spring-flowering hardy perennial with small mauve-pink to deep purple blooms in many-flowered umbels. Leaves more or less orbicular and lobed, coarse and sparsely hairy. Height 15–30cm (6–12in), spread 20–30cm (8–12in). An easy plant for the peat garden. Benefits from a little shade. Readily increased from seeds, the old plants being replaced periodically as they deteriorate rather quickly with age.

P. polyneura
Spring-flowering hardy perennial with pale rose, rich rose or purple blooms on slender flower stalks. Leaves triangular-ovate, green and downy. Height 15–30cm (6in–1ft), spread 20–30cm (8in–1ft). A good plant for the peat garden or richly organic soil at the waterside. Easily increased from seeds and like other species in this section best replaced every few years.

P. saxatilis
Short-lived spring and early summer flowering perennial with many-flowered umbels of rosy-violet or pinkish-mauve blooms. Rounded toothed and lobed leaves with long leaf stalks, green, soft and downy. Height 10–30cm (4in–1ft), spread 20–30cm (8in–1ft). An easy plant for a shady spot in a richly organic soil. Plants start to deteriorate after a

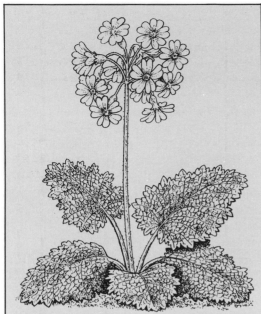

A good peat garden primula, P. polyneura is a hardy spring-flowering species with pale or rich rose or purple blooms on 15–30 cm (6–12 in) stems.

couple of years and need replacing. Easily raised from seeds sown during spring or summer.

P. sieboldii

Early summer flowering perennial with umbels of rounded white, pink or purple blooms with conspicuous eyes. Leaves ovate and toothed, soft green and finely downy. Height 15–20 cm (6–8 in), spread 15–20 cm (6–8 in). Gorgeous plants for a shady spot on the peat garden. Easily increased from seeds, but named cultivars by careful division in early spring or directly after flowering. In years gone by many cultivars were available. They now seem to have gone into decline and only old favourites like 'Wine Lady' are popularly available. Japanese cultivars are increasingly encountered, some with quite fascinating names in translation. At present these are mainly for the collector and enthusiast.

9. DENTICULATA

P. denticulata

(Drumstick primula) Very popular, vigorous spring-flowering perennial for a moist border or bog garden. Large heads of lilac, pink or purplish blooms crowded into globular heads on strong flower stalks during early spring. Leaves long, broad and coarse, often dusted with white or yellowish farina. Flower stems often coated in white or yellowish meal. Height 30–60 cm (1–2 ft), spread 30–45 cm (1–1½ ft). A native of the Himalayas and completely hardy. Easily increased from seeds sown during the spring or early summer, the seedlings flowering during their second year. Good forms can be reproduced from root cuttings taken during the dormant period.

P.d. var. alba

White Drumstick Primula Similar in almost every respect to *P. denticulata*, but producing heads of pure white flowers. Height 30–60 cm (1–2 ft), spread 30–45 cm (1–1½ ft). Extremely variable when raised from seed, some occasional purplish or lilac-flowered forms appearing. Vigour of the plant and density of the globular flower head are very variable among seed-raised plants. Reproduction by root cuttings of selected forms.

P.d. **var.** *cachemiriana* A vigorous variety of *P. denticulata* with fairly consistent purple-coloured blossoms. This is generally more robust than the ordinary species with slightly larger flower heads, and bold cabbagy foliage coated beneath with yellow meal. Height 30–60cm (1–2ft), spread 30–45cm (1–1½ft). Most plants are grown from seeds, but any particularly fine forms can be propagated from root cuttings during the dormant period.

P.d. 'Rosea' A rich crimson form identical to *P. denticulata* except in colour. Height 30–60cm (1–2ft), spread 30–45cm (1–1½ft). Best increased from root cuttings during the dormant period.

11. FARINOSAE

P. auriculata

A very variable hardy perennial, spring-flowering species from high mountain regions of the Caucasus and Turkey. Flowers small, lilac with a greenish throat. Leaves more or less lance-shaped in a neat rosette occasionally dusted with farina. Height 15–30cm (6in–1ft), spread 15–20cm (6–8in). Grows naturally by streamsides in mountainous areas. Prefers a gritty medium with a liberal quantity of sedge peat incorporated. Moist, but free-draining. Propagation is most easily undertaken by careful division of plants during early spring.

P. clarkei

A tiny spring-flowering hardy perennial primula producing rose-pink blooms with a yellow eye just above neat, tight clumps of fresh green foliage. Height 5–10cm (2–4in), spread 5–10cm (2–4in). A plant much loved by alpine house growers. Rarely succeeds in an open rock garden, but a little gem for pan culture under unheated glass. Cannot be grown successfully in a pan in the home. Propagate from seeds sown immediately after ripening or by careful division of the crowns after flowering.

P. darialica

This is usually the same plant as *P. frondosa* in the horticultural trade. Like that species, it is a spring-flowering hardy perennial for the alpine house or rock garden. Flowers lilac-rose to purplish-red with yellow eyes borne in small umbels on short stems above densely farinose foliage. The true *P. darialica* differs little from *P. frondosa*, so you only need consider growing one of them. Easily cultivated in a free-draining gritty compost. Reproduces freely from seeds sown during the spring or summer and can be increased by careful division in late spring or early summer, immediately after flowering.

P. farinosa

(Bird's eye primrose) One of the loveliest spring-flowering, hardy perennial primulas for scree or pan culture. Tubular shaped lilac-pink or occasionally white flowers with flared petals are borne in umbels on short stout stems. The neat rosettes of elliptic or ovate leaves are densely coated in white farina. Height 15–30cm (6in–1ft), spread 15cm (6in). Easily grown in an open position with good drainage, though it

Primula capitata *is a mid to late summer-flowering hardy perennial recommended for the peat garden or a shaded pocket of humus-rich soil on the rock garden.*

A gorgeous plant for a shady spot in the peat garden is early summer flowering Primula sieboldii *with very handsome toothed leaves and flowers in various shades, with conspicuous 'eyes'.*

does like access to a regular supply of moisture. In its natural habitat it will often colonize wet ground, but only where water drains away quickly and the soil never becomes stale or stagnant. These conditions can be difficult to contrive in the garden, so scree and pan culture are to be preferred. Well established plants can be carefully divided after flowering, but seeds are the most popular means of reproduction. Freshly sown seeds germinate freely.

P. frondosa
A spring-flowering hardy perennial primula for the rock garden or alpine house. An excellent small primula for the novice. Flowers lilac-rose to reddish-purple with yellow eyes are borne in umbels on short stems above rosettes of densely farinose foliage. Height 5–15cm (2–6in), spread 10–15cm (4–6in). This is one of the most amiable rock garden primulas, flourishing in most well drained soils. Unlike other alpine species it easily succumbs to winter wet. Easily divided after flowering. Rapidly reproduced from seeds, especially when gathered fresh. Seedlings often appear around mature plants and can be moved to other parts of the garden.

P. luteola
A most attractive hardy perennial spring-flowering primula which is more easily cultivated than its appearance would suggest. Lovely shallow funnel-shaped yellow flowers are carried in neat umbels on stout stems. The leaves, in rosettes, are more or less lance-shaped, bright green and glabrous.

Height 15–30cm (6in–1ft), spread 15–20cm (6–8in). A vigorous and easily grown species that benefits from cultivation in a pocket of free-draining gritty soil on the rock garden, but will adapt to border conditions if the soil is well-prepared. Grows well in full sun, but tolerates dappled shade. Seed raising is hazardous because germination is so erratic. Division after flowering is a simple means of reproduction from which the plant benefits. P. luteola begins to deteriorate if left undisturbed for three or four years.

P. mistassinica
A gorgeous little species from North America which is being increasingly found in cultivation. It is very similar in appearance to P. farinosa, but tends to have taller flower stems and smaller umbels of pale pink, lilac or occasionally white flowers. The green leaves are small, narrow and spatulate with a coat of white farina. Height 5–20cm (2–8in), spread 10–15cm (4–6in). Not as vigorous or easy to grow as P. farinosa, so most gardeners keep it in a pan in the alpine house. It requires a free-draining gritty medium, but a constant supply of moisture.

The plant can be propagated from seed sown during spring or early summer or by careful division after flowering during late spring. It can also be increased from adventitious buds which appear regularly on the roots and yield small plantlets. Remove these plantlets carefully and pot them as you would seedlings when they are large enough to handle.

P. modesta
One of the tiniest hardy spring-flowering perennial primulas for the alpine house. Flowers small, pinkish-purple in clustered heads on a short stem. Neat rosettes of bright green foliage covered with yellow meal. Height 5–10 cm (2–4 in), spread 10 cm (4 in). Demands a very gritty free-draining compost. Best grown fairly tightly in a small pan. Seed is difficult to come by, so divide carefully immediately after flowering.

P. rosea
An early spring-flowering perennial for a damp place in the garden. Beautiful glowing pink primrose-like blossoms appear among green leaves that are attractively flushed with copper or bronze. These spring tints are short-lived, the broadly lance-shaped leaves reverting to plain green. Height 10–15 cm (4–6 in), spread 15–20 cm (6–8 in). A wonderful plant for the stream or poolside. Flourishes anywhere, in sun or partial shade, provided the soil contains a liberal quantity of organic matter and is constantly moist. Easily increased from seeds, ideally sown directly they ripen. Plants produced in this manner are sometimes variable in colour and stature so .it is best to select the best forms for propagation by division. Divide them immediately after flowering.

P.r. 'Delight' An exciting cultivar with brilliant rose-pink blooms of superior quality and size. Height 10–15 cm (4–6 in), spread 15–20 cm (6–8 in). Must be increased by division.

P.r. 'Grandiflora' A larger flowered cultivar with similar rose-pink blooms and plain green leaves that have a hint of bronze in the spring. Not as consistent as 'Delight' because the plants offered in garden centres are often seed raised. The true 'Grandiflora' is increased by division to maintain consistency. It is prudent to mark the best plants in a group of 'Grandiflora' while in flower and reproduce them vegetatively. Height 10–15 cm (4–6 in), spread 15–20 cm (6–8 in).

P. scotica
One of Scotland's loveliest flowers. A delightful little character with dark lilac-purple blooms with a yellow throat in small umbels on short stout stems. The leaves form a clustered rosette, some upright, some flat, but all green and variously coated with farina. Height 5–10 cm (2–4 in), spread 10–15 cm (4–6 in). Many gardeners consider this to be a biennial, though it is technically a perennial. It tends to be short-lived in cultivation and within three or four years has disappeared. But it is so easily raised from seed that once you have it in the garden there is no reason why it should not be a permanent feature. Though relatively easy to grow, it profits from being grown in a trough or pan in a very gritty compost, but like most of the other primulas in this section needs constant access to moisture. Since this species is so short-lived, propagation is exclusively from seeds which usually germinate freely when sown fresh. Seeds kept from one year to the next

before sowing often take two years before germinating, even if subjected to a period of chilling.

P. yargongensis

Late spring flowering hardy perennial primula for a damp corner in the garden. Flowers tubular, bell-shaped, mauve, pink or purple with a white or cream eye in small umbels. Leaves ovate or elliptical, plain green. Height 10–30cm (4in–1ft), spread 15–30cm (6in–1ft). A native of wet streamsides in north-western Yunnan and south-eastern Tibet, this is an ideal plant for the poolside or peat garden. Benefits from a very damp peaty soil, but will rot off if forced to sit in permanent moisture. Easily raised from seeds sown immediately after they ripen. Plants set seeds freely provided they receive constant moisture.

P. yunnanensis

A lovely fragrant hardy primula with delicate bell-shaped blooms of rose-pink or lilac with conspicuous yellow eyes. Leaves soft, green, more or less elliptical and densely yellow farinose beneath. Height 10–30cm (4in–1ft), spread 15–20cm (6–8in). An excellent, but variable plant well suited to cultivation in the peat garden, or on the rock garden in a pocket in which the soil has been enriched with organic matter. Enjoys moist conditions, but resents sitting in the wet. Easily raised from seeds sown during spring or summer and by careful division in early spring just as it starts into growth, or immediately after flowering.

12.FLORIBUNDAE

P. floribunda

Tender perennial primula for the cool greenhouse or home. Spring flowering with distinctive golden-yellow narrowly cylindrical tubular flowers with attractively flared petals in small whorls. Ovate or elliptic, softly hairy green leaves. Height 15–20cm (6–8in), spread 15–20cm (6–8in). A splendid plant for the frost-free greenhouse. Grow in John Innes potting compost No. 2 and water regularly during the summer with a liquid feed. It is possible to divide plants, but much better results are obtained from seeds, especially if sown directly after harvesting. Use shallow pans and a soilless compost for seed raising, providing adequate light but shading from strong sun, which is likely to scorch tender seedlings.

P. × kewensis

The most important primula in this section. A spring-flowering tender perennial hybrid resulting from a union of *P. verticillata* and *P. floribunda*. *Primula × kewensis* is a widely grown and popular pot plant with bright yellow, fragrant, long bell-shaped flowers in neat whorls. Leaves obovate, green with a hint of farina. Height 30cm (1ft), spread 15–30cm (6in–1ft). This was originally a natural hybrid that was completely sterile. The plant in cultiva-

Opposite: *The drumstick primula,* P. denticulata, *is a very popular early spring-flowering species for a moist border or bog garden. Easily increased from seeds or root cuttings.*

tion is a fertile tetraploid form that comes true from seed. This is the most reliable way of propagating *P.* × *kewensis.* Grow in a frost-free greenhouse in John Innes potting compost No. 2.

P. verticillata
An early spring flowering tender perennial for the frost-free greenhouse. Long bell-shaped fragrant yellow blossoms in many-flowered whorls. Leaves lance-shaped, plain green and irregularly toothed. Height 20–30 cm (8 in–1 ft), spread 15–20 cm (6–8 in). An excellent spring flowering indoor plant that is easily raised from seed. Grow in John Innes potting compost No. 2.

14. MALACOIDES

P. malacoides
Tender winter-flowering perennial often treated as an annual. Flowers small, star-like, mauve, in many-flowered tiered whorls. Leaves small, rounded, lobed, soft green and downy. Height 20–30 cm (8 in–1 ft), spread 20–30 cm (8 in–1 ft). An easily grown seed-raised pot plant excellent for home or greenhouse. Grows successfully in soilless or soil-based compost, the best results usually being obtained by the home gardener in John Innes potting compost No. 2.

'**Ballerina**' A mixed strain with blooms of carmine-rose, lavender-mauve, white, coral and cherry.

'**Bright Eyes**' Large-flowered, early vigorous strain in a mixture of mauve, pink, rose, purple and white.

'**Carmine Rose**' A sturdy compact plant with carmine-rose flowers.

'**Coral Reef**' Deep coral-red flowers with a hint of orange. Well-proportioned umbels of blooms just above the neat compact mound of foliage.

'**Topsy**' Very free-flowering lavender-rose cultivar of neat, compact habit.

17. MUSCARIOIDES

P. muscarioides
Midsummer flowering hardy perennial with many-flowered heads of similar appearance to the spring-flowering bulb, the grape hyacinth. The individual flowers are very different, of course, being somewhat tubular, pendant and a deep purplish-blue. The leaves are obovate or elliptic, dull green and slightly downy. Height 30–45 cm (1–1½ ft), spread 15–30 cm (6 in–1 ft). Best accommodated in the peat garden, preferably with a little dappled shade. Easily raised from seeds sown as soon after harvesting as possible. Well established plants can be divided during early spring, as soon as signs of growth are detected.

P. viali
Late spring or early summer flowering hardy perennial, although some gardeners suggest that it is monocarpic and dies after flowering. This does happen, but it is usually the result of poor cultivation rather than the plant's inevitable behaviour. One of the most bizarre prim-

ulas sporting short spikes of flowers that look rather like miniature red-hot-pokers. Dense heads of small tubular flowers of red and bluish-purple. Small lance-shaped, soft, downy green leaves. Height 30–45 cm (1–1½ ft), spread 20–30 cm (8 in–1 ft). It will live comfortably in a damp border or bog garden, but this lovely little character is at its best in a peat garden. Easily raised from seed, preferably sown shortly after collection. Lifting and dividing plants is largely futile as almost half of them are likely to be lost during the operation. A regular supply of young seed-raised plants should always be maintained to counter the plant's tendency to fade out after flowering.

18. NIVALES

P. chionantha
Late spring flowering hardy perennial with fragrant ivory-white tubular bell-shaped blossoms in many-flowered umbels. Leaves green, broadly lance-shaped, toothed with yellow farina beneath. Height 45–75 cm (1½–2½ ft), spread 30–45 cm (1–1½ ft). A good plant for a well-drained spot on the peat garden. Enjoys a moist peaty soil and prefers a little dappled shade. Can be raised from seeds, preferably sown directly after harvesting. Established plants can be divided in early spring as soon as the shoots are seen poking through the soil.

P. macrophylla
An exceptional late spring flowering, hardy perennial plant for the enthusiast.

Fragrant purple or lilac-purple, rarely white, blooms in umbels of varying numbers. Leaves long and green, but with a white farina beneath. Height 15–30 cm (6 in–1 ft), spread 15–20 cm (6–8 in). A plant of mountain meadows that is at home in a constantly damp corner of the peat garden. Moisture is vital, but *P. macrophylla* objects to sitting in water where it will rot off during the winter. A sheet of glass raised on two bricks to protect the overwintering crowns from excessive damp is advisable. This primula can be readily raised from seed when available, but the best forms are grown from divisions taken in early spring.

P. purdomii
Late spring flowering hardy perennial with tubular bell-shaped flowers of lavender or purple on short flower stalks. Leaves lance-shaped, green, slightly farinose beneath. Height 15–20 cm (6–8 in), spread 15–20 cm (6–8 in). An unusual primula for the peat garden, enjoying a damp, but free-draining, organic medium, preferably with a little dappled shade.

P. sinoplantaginea
Late spring flowering short-lived hardy perennial. Deep purple, fragrant tubular blooms in small umbels. Leaves narrow, lance-shaped, smooth green with yellowish farina beneath. There are distinctive reddish scales at the base of the plant. Height 15–20 cm (6–8 in), spread 20–25 cm (8–10 in). It is always questionable whether plants offered by nurserymen under this name are really

Opposite: Primula denticulata *var.* alba, *the white drumstick primula, is best raised from root cuttings.*

Above: Primula farinosa, *the bird's eye primrose, is one of the loveliest spring-flowering species for pan culture or scree.*

the true species as it hybridizes so freely. Whether it is botanically correct or not matters little to the gardener, for it is a plant of great beauty. Established plants do not divide well, but seeds set freely and young plants are easily raised, especially if the seeds can be sown fresh. A plant for a moist, but free-draining spot at the waterside or a cool corner in the peat garden.

P. sinopurpurea

A delightful late spring flowering hardy perennial primula. Gorgeous violet-purple tubular blossoms in six- to twelve-flowered umbels. Broadly lance-shaped, plain green leaves with serrated edges, sparingly covered with farina beneath. Height 30–45 cm (1–1½ ft), spread 20–30 cm (8 in–1 ft). A more reliable and vigorous plant than *P. sino-plantaginea*, but of the same general aspect. Excellent for waterside planting, but equally happy in the peat garden. Readily raised from seed, especially if sown shortly after harvesting. Spring division is possible, but needs great care to ensure a reasonable percentage re-establishes.

Equally at home by the waterside or in a peat garden, Primula sinopurpurea opens its clusters of violet-purple blooms on 30–45 cm (1–1½ ft) stems in late spring.

19.OBCONICA

P. obconica

Winter and spring flowering perennial for the frost-free greenhouse or home. Large rounded funnel-shaped purplish or lilac blossoms each with a yellow eye borne in dense umbels. Leaves broad, ovate, mid-green, hairy, sometimes causing skin irritation. Height 15–30 cm (6 in–1 ft), spread 15–30 cm (6 in–1 ft).

An easily grown pot plant best raised from seed sown annually in late spring or early summer. Some gardeners keep their plants from year to year, but these tend to become fibrous and woody in the centre. Can be successfully raised in a soilless compost or John Innes potting compost No. 2.

There are many named varieties of *P. obconica*. Those available to the home gardener as seeds are mostly standard open-pollinated varieties. The modern F_1 hybrids are still in the hands of the commercial grower, though these are increasingly finding their way into florists' shops. They are well worth looking out for.

Modern F₁ Hybrids

'**Artemis**' Very free flowering white with a yellow eye.

'**Cluny**' Deep rose to carmine-magenta blossoms borne in profusion.

'**Isis**' Large blossoms of deep rose.

'**Salomon**' Creamy-orange flowers which intensify to deep orange with age.

'**Samba**' Unquestionably the best blue variety. Early, compact and free-flowering.

'**Senilis**' Rosy-mauve or wine-coloured blossoms in profusion. One of the most widely grown.

Standard Cultivars

'**Agnes**' Blue and white flowered cultivar. Outer parts of the petals vary from light to dark blue.

'**Bagatelle**' Warm salmon rose. Large flowered.

'**Concorde**' Clear, pure, mid-blue.

'**Fasbender Blue**' A reliable mid-blue variety.

'**Fasbender Red**' An old well-tried red-flowered cultivar.

'**Pink Velvet**' A bicolor, pink shading to white at the outer edges of the petals. The centre of the flower is almost crimson. Requires good light if its startling colour combination is to

develop to its best.

'**Royal Red**' Deep red with copper underlay.

'**Swift**' A frosty rose deepening to medium rose.

21.PETIOLARES

P. aureata
Spring-flowering hardy perennial with cream to yellow flowers in small umbels. Foliage mid-green with striking purplish mid-rib during summer, in winter a tight bud with conspicuous white meal. Height 10–15cm (4–6in), spread 10–15cm (4–6in). A plant for the alpine house or careful cultivation in the rock garden. Requires a richly organic, yet free-draining compost. A standard soil-less compost with up to 25% by volume of sharp grit would be reasonable, with plenty of drainage material in the bottom when grown in a pan. On a rock garden choose a cool pocket in a sheltered spot and provide adequate drainage. Propagation from seed is erratic and produces many inferior forms. Careful division after flowering is the most satisfactory method of reproduction.

P. bhutanica
Spring-flowering hardy perennial with pale blue blossoms, each with an ochre eye. Finely toothed obovate green leaves. Height 10–15cm (4–6in), spread 10–20cm (4–8in). This is considered by many to be a synonym of *P. whitei*, and most plants offered by nurserymen as *P. bhutanica* are thought to be this

species. A number of botanists argue that there are two distinct species in nature. *P. bhutanica* is cultivated as recommended for *P. whitei*.

P. edgeworthii

Spring-flowering hardy perennial with umbels of pale mauve blooms each with a white eye. Leaves broad, lustrous green, expanding and changing form considerably over the season after flowering. Height 10–15cm (4–6in),

Above: *An excellent small primula for the novice is* P. frondosa. *It is a spring-flowering species for the alpine house or rock garden. Often self-sows freely.*
Opposite: *For the stream- or pool-side,* Primula rosea *'Grandiflora' blooms in early spring and requires moist soil.*

spread 10–20cm (4–8in). A species for pan culture or careful cultivation in a cool partially shaded corner of the peat garden. Best grown under cold glass in a pan in a compost with plenty of organic

matter. John Innes No. 1 potting compost with up to 25% by volume of sedge peat added is suitable. Adequate drainage should be provided. Similar conditions should be provided when grown in a peat garden. A cloche or sheet of glass placed over the plant during winter helps to prevent the plants dying from excessive damp.

P. gracilipes
A late spring-flowering hardy perennial with a curious branched creeping rhizome. Flowers bright pink-purple with an orange-yellow eye surrounded by a narrow zone of white. The leaves are obovate or elliptic, green but often with a prominent red mid-rib. During winter the dormant crown and embryo flower buds are covered in farina. Height 10–15 cm (4–6 in), spread 10–20 cm (4–8 in). An excellent plant for a cool sheltered position on the rock garden in a well drained soil with plenty of organic matter incorporated. Propagate by careful division of established plants after flowering.

P. sonchifolia
Hardy spring-flowering perennial with dense umbels of blue-purple flowers each with a white eye margined with yellow. Short coarse green leaves at flowering time, which expand considerably during the summer. Height 20–30 cm (8 in–1 ft), spread 20–30 cm (8 in–1 ft). A plant for a moist position on the rock garden or in the peat garden. Must have a free-draining medium containing plenty of organic matter. In high rainfall areas it helps to protect the winter

crown of the plant against damage with a cloche or sheet of glass raised on a couple of bricks. Propagate this species by careful division immediately after flowering.

P. whitei
Spring-flowering hardy perennial with handsome umbels of blue or bluish-violet blooms with yellowish-green eyes. Leaves elliptic green with a sprinkling of white farina on each side. Height 10–15 cm (4–6 in), spread 10–20 cm (6–8 in). This plant is often considered synonymous with *P. bhutanica*. In cultivation both names usually apply to the same plant, but which one it really is, is often open to question. A first class plant for the peat garden, growing best when planted in the joints between peat blocks in an almost horizontal fashion. This ensures a richly organic medium from which excess moisture can drain away. Protect from dampness during winter with a sheet of glass raised on bricks or a cloche placed over the crowns.

26. SIKKIMENSIS

P. alpicola
(Moonlight primula) Early summer-flowering hardy perennial with pendant bell-shaped flowers varying in colour from yellow to white and purple carried on slender flower stems. Leaves small, rounded and bright green. Height 15–60 cm (6 in–2 ft), spread 15–30 cm (6 in–1 ft). A plant for a cool moist position, ideally suited to a partially shaded corner of the peat garden. Propagate by early spring division or seeds, preferably

sown shortly after harvesting.

P. alpicola alba A pure white-flowered form that comes almost 100% true from seed. Height 15–60 cm (6 in–2 ft), spread 15–30 cm (6 in–1 ft).

P. alpicola luna This is the beautiful sulphur-yellow flowered form. Like its white sister this comes almost completely true from seed. Height 15–60 cm (6 in–2 ft), spread 15–30 cm (6 in–1 ft).

P. alpicola violacea A purple to violet selection which is rather erratic from seed. Height 15–60 cm (6 in–2 ft), spread 15–30 cm (6 in–1 ft). To ensure plants of even stature and colour it must be propagated by division in early spring, just as the young shoots are emerging.

P. florindae
(Himalayan cowslip) A giant primula which flowers from midsummer into early autumn. Large heads of pendant sulphur-yellow blooms above broad, coarse green leaves. The whole plant has a musky aroma. Perfectly hardy and reliably perennial. Height 60–90 cm (2–3 ft), spread 30–60 cm (1–2 ft). A big bold bog plant that will grow in damp soil as well as standing water. A lovely streamside plant that fully tolerates the rising and falling of the water. Enjoys a deep cool soil with plenty of organic matter incorporated into it. Propagates freely from seed sown at any time during the spring or summer. Plants can be divided during early autumn when flowering is over.

P.f. 'Art Shades' This is a seed-raised selection with flowers varying from palest primrose and apricot to burnt orange. Handsome plants for adding colour variation during late summer. Height 60–90 cm (2–3 ft), spread 30–60 cm (1–2 ft). Though popularly seed raised, exceptionally fine colour forms can be reliably increased by division immediately after flowering.

P. ioessa
Spring-flowering hardy perennial with funnel-shaped pink or pinkish-mauve, occasionally white blooms in clustered heads. Leaves green, rounded and deeply toothed. Height 10–30 cm (4 in–1 ft), spread 15–30 cm (6 in–1 ft). Propagate from seeds sown as soon as gathered or by division of established plants during early spring. They can also be split up after flowering.

P. secundiflora
A charming hardy summer-flowering perennial with reddish-purple, pendant, funnel-shaped blooms borne in groups on slender flower stems. Leaves lance-shaped, green, often with a yellowish meal beneath in the spring. Height 30–45 cm (1–1½ ft), spread 30 cm (1 ft). Enjoys a little dappled shade and a cool, moist root run. Well suited to the peat garden or a bog garden provided it will not sit in water during the winter. Easily increased from seeds sown directly after harvesting or by dividing established plants in early spring.

P. sikkimensis
Hardy summer-flowering perennial with soft yellow, pendant, funnel-shaped

flowers on slender stems. Leaves coarse, green, rounded and shiny. Height 45–75cm (1½–2½ft), spread 30–45cm (1–1½ft). A true bog plant that revels in wet soils with a high organic matter content, preferably in partial shade. Easily increased by dividing established plants after flowering. Seeds sown in spring or early summer are usually quite successful, even if packeted the previous year.

P. waltoni

Late spring or early summer flowering hardy perennial with deep wine-purple, occasionally pink flowers in neat heads on strong wiry stems. Leaves lance-shaped, but rounded at tip, bright green. Height 30–45cm (1–1½ft), spread 20–30cm (8in–1ft). Needs a cool moist root run and appreciates dappled shade. Does not always divide freely, but is very easily raised from seed, especially if freshly gathered.

27.SINENSES

P. sinensis

A tender winter and spring flowering perennial usually treated as an annual. The species has large purplish-rose star-like flowers with yellow eyes. Leaves irregular, roughly heart-shaped, dull to olive-green, downy, often red beneath. Height 15–20cm (6–8in), spread 15–20cm (6–8in). A first class pot plant

Opposite: *Popular as a pot plant in the frost-free greenhouse or conservatory is* P. × kewensis. *This tender perennial hybrid blooms in the spring and is generally discarded after it has flowered.*

growing successfully in John Innes potting compost No. 2. Usually discarded after flowering as it rarely makes a second showy plant. Most seedsmen offer a selected strain in mixed colours.

28.SOLDANELLOIDEAE

P. nutans

Early summer flowering monocarpic species with bell-shaped pendant blooms of lavender or violet in conical heads on a stout stem above narrow green leaves. Height 60–90cm (2–3ft), spread 30–60cm (1–2ft). An excellent plant for the peat garden benefiting from a little dappled shade. Most seasons sets seed freely and so can be readily reproduced. Sow the seeds immediately after harvesting for the best results. This species is also grown under the name *P. flaccida.*

P. reidii

Late spring-flowering perennial with ivory-white pendant open bell-like blossoms in many-flowered heads. Leaves green, lance-shaped. Height 10–15cm (4–6in), spread 10–20cm (4–8in). A little gem for pan culture or a cool spot in the peat garden. Requires a richly organic, free-draining, but constantly moist compost. It can be a little tricky getting the watering just right, but for the primula enthusiast this is a delight.

30.VERNALES

P. amoena

Spring-flowering hardy perennial primula of similar disposition to *P. elatior.* Strong flower stems arise from among soft bright green leaves, often liberally coated beneath with white 'wool'. Each stem carries three or four bright violet-blue or lavender-blue blooms, occasionally white, each with a bright yellow eye. Height 10–15 cm (4–6 in), spread 15–20 cm (6–8 in). A most attractive plant for the alpine house, although practised primula growers seem to manage it outside quite well. Requires a rich organic compost. John Innes potting compost No. 2 with about 25% by volume of sedge peat is ideal. Free drainage is important, especially in the open. Propagate by regular division of established plants immediately after flowering. This helps the plants to maintain their vigour too. They should certainly not be left undisturbed for longer than three years. Seeds are usually set freely and can be used to provide a reserve of young plants. Take care to see that the promiscuous *P. amoena* does not cross with any other species.

P. elatior

(Oxlip) Spring-flowering hardy perennial with tubular sulphur-yellow blooms in neat umbels above soft green, somewhat downy foliage. Height 15–20 cm (6–8 in), spread 15–20 cm (6–8 in). A real favourite, much loved and increasingly cultivated for its ability to grow in the wild garden with little attention. Not to be confused with the hybrid between *P. veris* and *P. vulgaris* which it superficially resembles. Easily cultivated in any moist position in the garden in sun or partial shade. Grows well in a border, but is seen at its best

naturalised among grass where it will seed itself freely and form expanding colonies. Easily raised from seeds or by division immediately after flowering.

A number of sub-species are sometimes found in cultivation and often find their way on to the seed lists of specialist societies. These are often designated as species, but the name given is really the sub-species name for a part of *P. elatior*. Heights and spread are the same as for the species.

P.e. subsp. *leucophylla* A version of the common oxlip native to limestone areas. Differs mainly in the leaves which are grey tomentose beneath and narrow gradually to the leaf stalk.

P.e. subsp. *lofthousei* This is undoubtedly the best kind of oxlip for the gardener. A native of southern Spain, this has much darker yellow, smaller saucer-shaped blossoms in many-flowered umbels.

P.e. subsp. *pallasii* An interesting variation with fewer flowered umbels and narrowed leaf blades which are slightly toothed and glabrous.

P. juliae

Hardy spring-flowering perennial with deep magenta tubular blossoms with cleft ends to the petals. Flowers are produced singly from the centre of a creeping stoloniferous plant with roundish serrated green leaves with a purplish cast. Height 5–10 cm (2–4 in), spread 10–15 cm (4–6 in). A first class, easily grown creeping plant for a moist position. Excellent beside a pool, at the streamside or for a damp spot in a herbaceous border. Readily increased by division of the creeping rootstock. A parent of the important Juliana hybrids. These are referred to botanically as *P. × juliana* or *P. × pruhoniciana*.

P. × pruhoniciana – JULIANA HYBRIDS

These are derived directly from *P. juliae* and various coloured forms of *P. vulgaris*. They have been further developed by inter-crossing the hybrids. One of the most important groups of spring-flowering plants. They are divided into primrose or polyanthus habit and the varieties are available in almost every flower colour. Leaves mostly green or with a purplish cast, ovate to lanceolate and rather coarse. Height 5–15 cm (2–6 in), spread 10–15 cm (4–6 in). All are easily increased by division immediately after flowering and, like *P. juliae*, benefit from moist conditions.

Juliana hybrids with a primrose or semi-polyanthus habit.

'Afterglow' Rusty orange-red with a distinctive eye.

'Apple Blossom' A very popular pale apple-blossom pink.

'Belvedere' Large-flowered lilac cultivar, an old and well-tried favourite.

'Berry Green' Crimson blossoms with a bright yellow eye.

'Blue Horizon' A vigorous and very free-flowering plant with greyish-blue blooms with a yellow eye.

'Bunty' Deep blue flowers shading to purple-red towards the centres and with yellow eyes.

'Cherry' As its name suggests this is a bright cherry-red.

'Craddock White' A bold white flushed with cream, having a striking yellow eye.

'Crimson Queen' One of the largest red-flowered hybrids.

'Dinah' Burgundy-red blooms with olive-green eyes.

'E. R. Janes' An exceptional and popular cultivar with salmon-pink flowers flushed with orange or peach. Easily grown and very free-flowering.

'Gloria' An interesting scarlet-flowered cultivar with whitish markings on the inside edge of the petals. Bright yellow eye.

'Icombe Hybrid' A very vigorous hybrid with large rosy-mauve blooms with distinctive white eyes.

'Jewel' One of the most compact hybrids with deep reddish-purple flowers.

'Lilac Time' A pale rosy-lilac cultivar which produces its flowers rather like a polyanthus, but on short stout stems.

'Lingwood Beauty' Gorgeous red flowers with deep orange eyes. An excellent plant reproducing freely and quickly making sizeable carpets of growth. A favourite and very weatherproof cultivar.

'Morton Hybrid' A very squat plant well suited to a windowbox or sink garden. Bright red blooms, each with a large yellow eye.

'Mrs MacGillavry' Violet-mauve flowers in abundance. A very popular and freely available plant.

'Purple Cushion' A gorgeous purple-red cultivar with handsome purplish leaves.

'Purple Splendour' Large reddish-purple flowers with pale yellow eyes.

'Queen of the Whites' Probably the finest of all the clear whites. Usually spoiled by excessively wet weather, so may merit temporary protection from rain with a cloche in the spring.

'Romeo' A very vigorous large violet-flowered variety.

'Wanda' Unquestionably the best known of all the 'Juliana' hybrids. Very widely grown and having fine blooms of purple-red among green leaves with a hint of purple. A very resilient plant that should be in everyone's garden.

'Wisley Crimson' Very large purple-red flowers among fine deep bronze-green foliage.

Juliana hybrids with a polyanthus habit.

'Barrowley Gem' An early-flowering yellow cultivar with green shading.

'Ideal' A popular purple-flowered cultivar with a yellow eye.

'Kinlough Beauty' Soft salmon-pink blooms with a cream stripe down each petal.

'Lady Greer' One of the smallest flowering varieties, but having a charm of its own. Pale primrose-yellow blooms.

'Tawny Port' An established favourite, rather like a small polyanthus. Port-wine coloured blooms are produced above leaves with a strong reddish cast. Has one of the longest flowering periods.

'The Bride' A fine, cool icy-white cultivar.

P. megaseaefolia
One of the earliest flowering hardy perennial primulas with magenta-rose to rosy-pink blooms each with a whitish eye and yellow throat. These are carried on short stems in an umbel of similar appearance to a polyanthus. Leaves rounded, green, somewhat hairy. The plant gets its name from the shape of the leaves, which are almost identical to those of *Megasea*, the old name for *Bergenia*. Height 10–15cm (4–6in), spread 15–20cm (6–8in). Although it appears to be a tough and resilient plant, *P. megaseaefolia* is very prone to damping off in excessively wet weather. A good plant for the frame or alpine house, but if grown outside must have the winter protection of a cloche or

Opposite: Primula malacoides *is a tender winter- or spring-flowering perennial which is generally discarded after flowering. Ideal for a cool greenhouse or conservatory.* 'Bright Eyes' *is one of the many strains available.*

sheet of glass. Demands a free-draining compost with a reasonable quantity of organic matter mixed into it. John Innes potting compost No. 2 with about 25% by volume of an equal parts by volume sedge peat and sharp grit mix suits them well. Seeds are rarely offered, so propagate by careful division immediately flowering is over.

P. veris

(Cowslip) One of the commonest and most widespread primulas. Hardy spring-flowering perennial with tubular soft yellow blooms with distinctive orange patches in the throat. Leaves ovate, mid-green, soft, more or less downy. Flowers and foliage fragrant. Height 10–20cm (4–8in), spread 15–20cm (6–8in). A meadowland plant that grows happily naturalised in grass where it can be allowed to form expanding colonies. An interesting addition to the mixed border, but equally at home in the alpine house.

The cowslip requires a richly organic, moisture-retentive soil. Will grow in sun or partial shade. Easily increased from seed, though like other species in this section the cowslip is rather promiscuous and the progeny may not be wholly true. Division immediately after flowering is the best way of ensuring consistent quality. Several sub-species are found in cultivation, usually offered as species under their sub-species name. Heights and spreads of those listed below are the same as for the species.

P.v. subsp. *canescens* A plant for the enthusiast, this differs little except in foliage which is grey-tomentose. Flowers yellow.

P.v. subsp. *columnae* More open yellow flowers with longer tubes. Like a slightly enlarged version of the ordinary species.

P.v. subsp. *macrocalyx* From the gardener's point of view the best cowslip of all, as the tubes and petals are much larger and showy. The sub-species is variable – mostly yellow, but other coloured forms have been encountered. Packaged seeds of this sub-species usually yield cowslips varying in colour from yellow through ochre, to rust and red. In such commercial stocks, which are doubtless re-selected and possibly hybridised, occasional plants are produced with much reduced petals. Discard these as they are undesirable characteristics which seem to be dominant in any breeding programme.

P. vulgaris

(Primrose) A widely distributed and much-loved spring-flowering hardy perennial primula with soft yellow blooms with an orange or gold central patch. Leaves soft green, long, lance-shaped in neat clumps. Height 5–15cm (2–6in), spread 15–30cm (6in–1ft). An easily grown plant which naturalises well in grass. The common species is most adaptable to border conditions in any moist soil. Easily raised from seeds sown directly they are ripe, or by division immediately after flowering. The sub-species of *P. vulgaris* are of the same height and spread.

P.v. subsp. *balearica* White fragrant flowers, leaves with leaf-stalks longer than the blades.

P.v. subsp. *sibthorpii* A widely cultivated plant with rose, lilac or crimson blossoms. The one popularly grown under this name has lilac-pink flowers, but it has been involved in the development of many different primrose and polyanthus cultivars and often appears as an unnamed hybrid masquerading as subsp. *sibthorpii*. Most of the single-coloured primroses grown in gardens will be found under *P.* × *pruhoniciana* – Juliana hybrids. Primroses raised from seeds for the pot plant trade are dealt with separately. See Greenhouse Primroses, p. 82.

Double primroses These are fully double-flowered cultivars reproduced from division. They are used for garden decoration rather than pot culture.

'Alba Plena' A fully double white which requires regular division every couple of years to ensure that it retains its vigour and character.

'Bon Accord Beauty' Large purple-blue blooms edged with white.

'Bon Accord Blue' A fully double, large-flowered blue cultivar.

'Bon Accord Cerise' Double cerise-pink.

'Bon Accord Lilac' One of the most frequently encountered cultivars. Lovely soft lilac.

'Chevithorne Pink' A widely grown soft shell-pink.

'Crimson King' A rich velvety red double with flowers borne in a similar fashion to a polyanthus.

'Marie Crousse' Large pinkish-purple flowers with a tracery of white. A first class vigorous cultivar for the open garden, but I have had it flourishing in the alpine house where it makes a magnificent potful.

'Quaker's Bonnet' An old double lilac cultivar.

'Viride Flore' A doubtful name for what gardeners call the Old Double Green. A curio which is still commercially available. Not a plant of great beauty.

Double primroses enjoy cool damp conditions and benefit from partial shade. The soil should be moist and rich in organic matter. The secret of success with double primroses is to lift and divide them regularly. Never leave a clump undisturbed for more than two years.

Jack-in-the-greens This is a strange group of primroses whose origins date right back to the sixteenth century. The flowers are single, but behind them is a ruff of small leaves. Relatively few remain in cultivation, but those that do have a popular following. Treat exactly as you would double primroses. Some have their flowers arranged on a stem like a polyanthus, others produce their flowers from among the leaves like the common primrose.

'Donegal Danny' Pink flowers of polyanthus habit.

With flowers that look like miniature red-hot pokers, Primula viali *flowers in spring or early summer and is suited to a damp border, bog garden or peat garden.*

The Himalayan cowslip, Primula florindae, *is a giant species which flowers from mid-summer into early autumn. An excellent choice for bog garden or streamside.*

'Eldorado' Bright yellow blooms of polyanthus habit.

'Maid Marion' Large golden-yellow blooms with darker centres. Of primrose habit.

'Robin Hood' Startling red flowers edged with white. Of polyanthus habit.

'Tipperary Purple' Rich mauve-purple flowers of primrose habit.

Hose-in-hose primroses These are most bizarre flowers, each bloom having two rows of petals, giving the illusion of two flowers, one inside the other. Most are scented. They need the same cultivation as double primroses and Jack-in-the-greens.

'Aberdeen Yellow' An old and popular cultivar with small yellow flowers.

'Castle Howard' Soft primrose yellow.

'Irish Molly' An interesting mauve-pink cultivar sometimes listed as 'Lady Molly'.

'Lady Lettuce' Apricot-yellow blossoms flushed with pink.

'Ruddigore' Deep crimson red.

'Sparkle' One of the finest crimsons.

□ GREENHOUSE PRIMROSES

This is not a proper botanical designation, but is the gardener's name for the seed-raised strains developed for the pot plant market as distinct from those used for general garden decoration. All those grown by the glasshouse industry are perfectly hardy, but do not perform so well outdoors as under glass and often lead to disappointment. While the garden is a suitable repository for primroses that have flowered indoors, expectations should not be high for their continued usefulness. Pot grown primroses are easily grown in a standard soilless compost or a soil-based compost like John Innes potting compost No. 2.

Roggli Wonder primroses These are a mixed strain for producing early blooms. While all primroses will tolerate cold and variable temperatures, this strain is the best for pot plant culture in nothing more luxurious than a cold frame. Indeed, they do not respond well to high temperatures without distorting. A mixture which contains most popular colours. Plants of neat, compact habit.

Ducat primroses Another early strain which is becoming very popular with florists. This responds to more heat than the Roggli Primroses and is more difficult for the home gardener to raise successfully. It has much less polyanthus blood in it than other primroses and is therefore much more typical of the traditional primrose form.

Saga primroses Seeds of these wonderful primroses are not yet widely available to the home gardener, but the plants are taking the florists by storm. They can be forced into flower very early in commercial conditions and produce the most magnificent heads of flowers in single colours with bright yellow eyes. The cultivars are named merely by colour, for example, 'Saga Pink', 'Saga Blue' and 'Saga White'.

These are well worth acquiring as pot grown plants. With careful division after flowering they can be grown on from year to year.

Europa primroses An extraordinary range of primroses with brilliantly coloured flowers the size of Pacific Giant polyanthus. Seeds of these are not generally available, but the plants are widely sold in shops and garden centres in the spring. The separate cultivars are named 'Europa Blue', 'Europa Cream', 'Europa Gold' and so on. An excellent mid-season strain.

Finesse primroses These are neat compact-growing plants with flowers in a range of colours extending from blue, mauve and rose to scarlet. Almost all have a conspicuous white or gold edge to the petals. Mid-season flowering and tolerant of quite cool conditions.

Biedermeier primroses The most traditional and easily obtainable of the mid-season pot-grown primroses. A gorgeous array of colours on small compact plants. A reliable strain.

Laurin primroses Large-flowered, compact plants which are invaluable for extending the flowering period. In their myriad colours these last as long as the regular outdoor strains in the garden.

Nordic primroses Another late-flowered mixed strain, but one which can be grown in cool frame conditions. They do not respond well to forcing.

□ **MINIATURE PRIMROSES**
In recent years strains of miniature primroses have been bred specifically for growing in bowls or tubs. Although hardy when well established, these are usually treated as pot plants.

'Fairy Lights' A completely new strain with a colour range from white, yellow, pink and red to blue, lilac and violet.

'Juliet Cheerleader' A mixed strain in pastel shades. Pale to deep pink with conspicuous yellow eyes.

□ **POLYANTHUS**
This is a complex cross between a number of species within the *Vernales* section. It serves little purpose for us to try and analyse the parentage of the modern polyanthus, or indeed those that were developed on the way and are still maintained in cultivation. The groupings of polyanthus and polyanthus-type primulas here often have little botanical foundation, but they are those popularly used by gardeners.

Gold laced polyanthus These are beautiful old florists' varieties with a dark ground colour and gold lacing around the edges of the petals. Nowadays those that are popularly available are often of seedling origin, but a few old named varieties can still be found.

'Black and Gold' Dark velvety deep purple-black with a yellow eye and gold lacing.

'George the Fourth' Rich crimson with a bright yellow eye. Fine gold lacing.

'Prince of Orange' Dark ground colour with a yellow eye and regular lacing.

'Prince Regent' Dark reddish brown ground with a bright eye and regular lacing.

Gold laced polyanthus should be regularly lifted and divided to maintain their vigour. This is also the best means of propagation.

Barnhaven polyanthus These are famous strains of polyanthus which have

Primula obconica: *a winter/spring-flowering perennial for the frost-free greenhouse or home. Usually raised annually from seed and discarded after flowering.*

been cherished by gardeners over the past few decades. They were so good of their period that a mystique grew up around them. They are still very good and can still be obtained commercially. However, though it may be unpopular, I

For the rock garden or alpine house is the Petiolares primula, P. aureata. It flowers in the spring and requires a richly organic yet free-draining compost.

have to say that some of the modern commercial varieties now offered are far superior. I am sure that Barnhaven 'Silver Dollar' polyanthus – so named because each flower was the size of a silver dollar – will continue to be grown for many years and deservedly so, but they are not now the be all and end all of modern polyanthus.

Cowichan polyanthus This is one Barnhaven polyanthus which I would still grow. A strain developed from a chance seedling find in Cowichan, British Columbia, Canada. The original plant was red with no eye. Although it was self-sterile it was used as a pollen parent and from this a lovely strain of mixed-colour polyanthus has been raised without eyes and with handsome reddish-bronze foliage.

Pacific Giants A very fine strain of polyanthus that has been with us for some years. Very good garden plants with bold trusses of brilliantly coloured weatherproof blooms on stout flower stems. Large, vigorous green leaves. An excellent strain for the beginner. Easily raised from seeds. There is now a dwarf version of Pacific Giants called 'Dwarf Pacific Mixed' which is used for pot plant production and for sale in garden centres as plants for windowboxes and containers. Seeds of this do not yet seem to be available to the home gardener.

Crescendo hybrids One of the best modern bedding strains of polyanthus. Very winter hardy and reliable. A wide range of brilliantly coloured flowers on stout flower stems, so strong and vigorous that they can be used for cut flowers. Mixed Crescendo hybrids are available to the home gardener as seeds and plants are often available in separate colours in garden centres. The orange and red shades are particularly fine. Some commercial growers offer this as a house plant, but it has to be grown cool if it is to keep its neat habit.

Renown Brilliant hybrids Compact plants for spring bedding. A wide colour range with weatherproof blooms. Very leafy and with exceptionally strong flower stems.

Southbank polyanthus If you prefer polyanthus of the size and habit typical of those seen in cottage gardens, then this is for you. Medium-sized blooms.

Garryarde polyanthus These are an old group of polyanthus of mixed and varied parentage which are seeing a revival. They have a mixture of habits and forms which vary from the polyanthus type to the primrose. All enjoy moist conditions and associate well with the 'Juliana' primrose hybrids. Propagation is by division, which should be done regularly to maintain vigour. When listed in catalogues most will be prefixed 'Garryarde'. They have primrose-like leaves with a bronze cast.

'Buckland Red' A rich velvety red.

'Guinevere' The most popular and widely cultivated. Large pink blooms.

'Sir Galahad' Large white blossoms with frilled petals.

'The Grail' Bright brick-red flowers with a large yellow eye.

'Victory' Beautiful rich purple.

Hose-in-hose polyanthus
There are hose-in-hose polyanthus, as well as primroses, with two rows of petals giving the illusion of one flower inside another.

'Brimstone' Sulphur-yellow blooms.

'Goldilocks' Flowers of deep rich yellow.

'Irish Sparkler' Bright red blossoms.

'Old Spotted Hose' Crimson red, each petal spotted towards the edge.

'Windlestraw' Pale primrose-yellow flowers.

☐ **GREENHOUSE POLYANTHUS**
This is not a proper botanical designation, but a grouping gardeners use to distinguish seed-raised strains developed for the pot plant market from those used for spring bedding in the garden. All those grown as pot plants are hardy, but often do not perform as well outdoors as under glass. Pot grown polyanthus are easily grown in standard soilless compost or a soil-based compost like John Innes potting compost No. 2.

Presto polyanthus These are very early flowered polyanthus of unnaturally dwarf habit, but perfect for cultivation as pot plants. Seeds are not generally available to the home gardener, but plants are sold in large numbers through garden centres. A superb strain in almost every colour imaginable. Plants are often sold to colour, specific coloured variations coming absolutely true from seed.

Largo polyanthus Mid to late season flowering polyanthus grown as pot plants. Neat compact plants in a wide colour range. Seeds are not generally available to the home gardener, but plants are produced in large quantities and are available in florists' shops and garden centres.

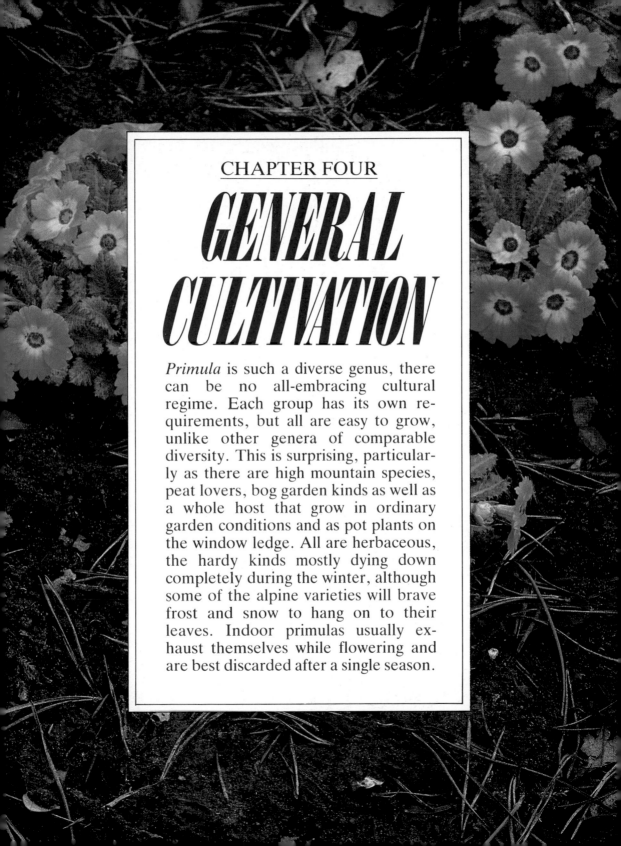

CHAPTER FOUR

GENERAL CULTIVATION

Primula is such a diverse genus, there can be no all-embracing cultural regime. Each group has its own requirements, but all are easy to grow, unlike other genera of comparable diversity. This is surprising, particularly as there are high mountain species, peat lovers, bog garden kinds as well as a whole host that grow in ordinary garden conditions and as pot plants on the window ledge. All are herbaceous, the hardy kinds mostly dying down completely during the winter, although some of the alpine varieties will brave frost and snow to hang on to their leaves. Indoor primulas usually exhaust themselves while flowering and are best discarded after a single season.

INDOOR PRIMULAS

By these we mean the popular house-plant varieties of *Primula obconica* and *P. malacoides*, though the coloured primroses widely sold as pot plants on Mother's Day must not be ignored. All are raised from seed, pricked out into trays, then potted individually. The compost used can have a marked effect upon the plant's development and longevity, so it is vital to get this right from the outset. The kind of compost and its nutrient value have a considerable bearing upon the plant's growth and also upon the gardener's ability to water it satisfactorily.

More indoor primulas are killed by faulty watering than by any other cause, whether over-watering, under-watering or just irregular watering. While each is linked with misuse of the watering can, it is more often the result of using an inappropriate compost, or the lack of understanding of an appropriate one. The subtleties of various composts are of little account here. It is sufficient to understand that there are two main types – soilless and soil-based. Those known as soilless consist of peat, or peat and sand, with a balance of nutrients added. The soil-based kinds are dominated by the John Innes composts, the most consistent and reliable of those with a significant soil content. John Innes composts are available in several forms, but generally consist of peat, sand and loam – a somewhat mythical soil which is ideally fibrous and midway between sandy soil and heavy clay. Soilless and soil-based composts can serve the same purpose, but require slightly different treatment to achieve success.

Soilless composts are quick-acting, light, and easy to handle, young plants advancing quickly in them. These composts are also useful for longer term primula culture, for primulas require regular repotting, so this medium should not significantly deteriorate in the short period between pottings. But remember that the peat in such compost is an organic material that eventually decomposes, a process hastened by regular feeding with liquid house plant feeds. As peat forms a substantial part if not all of the compost, decomposition can create problems if repotting is not regular. Apart from the hostile airless conditions which develop in the compost and restrict proper root development, mosses and liverworts invade the surface and sciarid flies move in. These, together with the general anaerobic conditions of the compost can quickly kill a large proportion of essential root hairs.

Soil-based composts, on the other hand, do not yield such rapid results, but are more stable. Watering is easier to get right. The soil also serves as a buffer against the breakdown of structure caused when soilless compost is regularly liquid-fed. The soil allows for a greater margin of error when watering, as surplus moisture can percolate through the compost more rapidly than if it were entirely of peat. It is the peat's excessive moisture-holding capacity, together with the difficulty of wetting it when it has been allowed to dry out, that make soilless composts more difficult to manage. Ideally the indoor gardener should use John Innes No. 2 potting compost for all actively growing plants in the home and use clay pots whenever possible. Not that these are any better for the plants, but the moisture content of the compost in a clay pot can be more easily assessed than that in a plastic pot. Take a short length of dowel rod or a stick of similar substance and tap the pot. If a dull thud results the compost is damp, if a ringing sound is heard, it is dry.

When the most suitable compost is used, watering is less hazardous and primulas have an excellent chance of prospering if given warm humid conditions with plenty of light. Little else need be done to keep them in good order provided they are regularly sprayed against the inevitable aphids. The relatively few horrors that trouble the indoor primula grower are discussed later, but most common pests likely to appear in the home can be controlled with an aerosol house plant insecticide spray. Fungal diseases are rarely troublesome. Mildew occasionally manifests itself as a white deposit on the foliage. If it should appear, spray with a systemic fungicide based on benomyl.

Light is important for the successful cultivation of primulas. In most homes the ratio of warmth to light, especially during spring and autumn, is completely out of proportion. High temperatures coupled with poor light intensity result in drawn plants with pale foliage and sparse blossoms. Primulas generally dislike really high temperatures, so treat them as cool house plants. Good light does not mean that they enjoy standing in a hot, sunny, south-facing window, although sunshine for a good part of the day is desirable. The full glare of the sun through a window is likely to scorch and dry the foliage and spot it if droplets of water remain on the leaves and the sun shines through. Drops of water act like magnifying glasses and bright sun can scorch the tissue beneath, often followed by secondary fungal infections that give the plants an unpleasant spotty appearance. Foliage damaged in this way survives rather than recovers, remaining unsightly until the end of the season. This trouble can occur even when the sun is only moderately fierce, so when watering take care not to splash the foliage.

Like all indoor plants primulas benefit from feeding, especially when heavily laden with blooms. Use a proprietary house plant feed, commencing when the first buds are showing colour and continue at regular intervals until the plant starts to grow more slowly in late summer. This only applies to plants intended to be kept for another year. *P. sinensis* and *P. malacoides* are often completely exhausted after flowering and best discarded, but *P. obconica* always seems prepared to go on indefinitely, though flower quality deteriorates after a few years.

The coloured primroses available in spring are not quite the same as the other indoor primulas. Though perfectly hardy, these modern florist's hybrids are not good garden plants. They are in-between plants that require varied cul-

tivation, spending the autumn, winter and early spring indoors, but the summer in a shaded place in the open. The pots should be plunged when the plants are put out so that their roots are kept cool. They can be plunged to their rims in the garden, or else in a deep box of peat or ashes. Protect from slugs with slug pellets, keep well watered, then in early autumn clean them up, repot and take indoors. If they have made good growth in the garden during the summer they can often be divided when being repotted.

Repotting often causes some consternation to newcomers to gardening, as many are unsure when to perform it. It is obviously better to repot a plant just before it needs it, but a newcomer may have difficulty in recognizing just when that is. Consequently plants are often allowed to go beyond that point and start to deteriorate. Pale foliage and a gaunt appearance, often accompanied by sparse flowers are characteristic of a primula that needs repotting. The rootball will be hard and congested, often with roots pushing out through the pot's drainage holes. The compost surface will probably look stale, with mosses or liverworts present too. Most healthy primulas will rapidly recover from their ordeal when repotted but it is better to catch them before they decline so that strong healthy growth can continue unchecked.

Opposite: Primula sinensis *is a tender winter- and spring-flowering perennial usually treated as an annual. It is a first-class pot plant. The variety 'Dazzler' is shown here.*

Do not be frightened to turn a plant out of its pot while it is actively growing and inspect its rootball. There is no need to pull it about, but regular inspection will show whether everything is in good order. Do not be over concerned about the concentration of roots round the sides of the pot, for it is quite natural for them to gravitate there. It does not necessarily indicate that the plant must be repotted. Similarly roots that push through drainage holes may not indicate congestion within the pot, for if the pot has been stood on a tray of moist gravel it is quite normal for roots to probe around outside. The best way to tell whether a plant needs repotting is to pinch the rootball with your fingers. If there is any give in the compost it shows that the pot-ball has not been completely ramified by roots, so nothing need be done yet. However, if the rootball feels hard and solid, repotting is clearly urgent.

When repotting ensure that there is adequate drainage in the bottom of the pot. A generous layer of small pea gravel or pieces of broken clay pot over the drainage holes should be adequate if one of the standard brands of potting compost is used. Remove the plant to be repotted from its pot by inverting it and giving the rim a sharp downward blow on the potting bench. If the rootball is very congested, run a knife down it in two or three places to allow fresh roots to break out. If the rootball is not tightly congested and just firm, then leave it undisturbed. Position the plant in the centre of its new pot and gently firm down the compost with your fingers.

Potting-on: (a) Well-rooted plant ready for move into larger pot. (b) To remove plant from its pot, hold it upside down with its base between the fingers, tap rim of inverted pot on edge of potting bench to dislodge root ball. Compost must be moist. (c) Plant set in its new larger pot with fresh compost. Do not over-pot.

Lightly top off the compost, allowing a 1 cm (⅓ in) space above the compost for watering. If you are using a John Innes potting compost, firm it before watering. When a soilless kind is used fill the pot with compost to the rim then water. This will settle it down so there is no need to firm it with your fingers. Tightly-packed soilless compost is devoid of air and virtually impervious to water, especially if its surface is allowed to dry out.

We have so far only considered indoor primulas grown in the home, but some gardeners are able to grow them in a greenhouse. This provides better scope for achieving perfection, even with the constraints imposed by neighbouring plants which may not enjoy exactly the same conditions. Primulas are easy going and the kind of atmosphere appreciated by most popular conservatory plants will suit them well.

The need for regular watering may seem obvious, but it is surprising how many gardeners only water when they have to, to the detriment of the plants. Good stable growth can never be achieved by irregular watering. A humid atmosphere is beneficial, especially during the summer, so spray the path and gravel under the benches regularly. Avoid getting water on to the plants' foliage during hot sunny weather, and provide the greenhouse with some kind of shading during late spring and summer.

Ventilate freely during warm weather and moderately during cooler periods. A free circulation of air among the plants helps reduce fungal diseases like botrytis. During the duller days of winter and early spring ensure that maximum light is admitted to the greenhouse. Clean the glass in the autumn and again during early spring. Cleanliness throughout the structure is important if pests and diseases are to be successfully controlled. Thoroughly cleanse rafters, brickwork and other fixtures with a solution of potassium permanganate and water each winter.

POLYANTHUS AND PRIMROSES

The primrose, and to a lesser extent the polyanthus, are now used as low energy crops by commercial glasshouse pot plant growers. Our view of their usefulness has been considerably widened and characteristics have been developed in some varieties that make them much more a florist's plant than a garden flower. We have just discussed how to look after these plants indoors, but growing them to the stage where they are potted demands exactly the same cultural regime as polyanthus and primroses used for spring bedding. Indeed, one or two of the modern cultivars developed for the pot plant market can equally well be grown as bedding plants, but especially for tubs, containers and window boxes. Let us consider these

very popular garden features first.

Any windowbox or planter must have a minimum depth of 15 cm (6 in) for proper root development – preferably deeper. With this in mind, carefully consider any urn or similar container with a shallow area around the rim before purchase, for if the outer planting cannot penetrate into the centre of the container it will frequently dry out and the plants will be permanently under stress. Drainage must also be provided, preferably by holes in the base of the container or windowbox, though a generous layer of gravel in the bottom in containers without holes will often suffice if watering is done carefully. Tubs, planters and windowboxes need careful attention to keep them looking their best, but the success of any container is based initially upon the plants and compost it contains. The wrong compost and unsuitable plants make it almost impossible for even the most skilful gardener to produce an attractive display. Composts should be moisture retentive, yet free-draining. For tubs and planters they are best based on the John Innes formula with at least one third by volume of peat added.

Windowboxes are different, for it is obviously better if they are light in weight. Soilless composts are therefore preferable. However, all containers demand regular watering, particularly where polyanthus and primroses are grown. It is essential to soak tubs, planters and windowboxes thoroughly and regularly during dry weather, rather than sprinkling them frequently with water. If they are well drained, surplus

Coloured forms of Primula vulgaris
*(primrose)are ideal companions for small
spring-flowering bulbs such as blue muscari.*

water will quickly run away, so damp-
ing-off problems will not arise. When
watering has been too light, the plants
produce roots just beneath the soil
surface where they are vulnerable to sun
and particularly drying spring winds. It
is not sufficient to water only enough to
darken the surface of the compost.
Surplus water must be seen running out
of the drainage holes at the base.

Whether grown in containers or the
open ground, the popular hybrid strains
of polyanthus and primroses require
similar treatment. They are sown in
trays in a cold frame during late spring
and by mid-summer are pricked out into
trays and by autumn ready for planting
out. Some gardeners move the plants
from trays as soon as they have developed sufficiently, continuing to grow
them on in pots until planting time. This
is more labour intensive, but does produce better quality plants, essential for
high profile plantings in containers and
windowboxes. Plants for massed bedding can be somewhat smaller and not
of such high quality, yet not be obvious
to the casual observer. Indeed, the extra
labour involved in producing pot grown
plants for more extensive bedding is not
justified.

Many gardeners have polyanthus
plants that they have used for spring
bedding for many years. Not the same
individual plants, but plants of the same
stock, for polyanthus are easily divided
after flowering. Immediately they have

finished flowering they are lifted from the beds and moved to temporary summer quarters. The usual practice is to divide the plants and line them out in nursery rows, where they are kept watered, then replanted in the beds once the summer bedding has faded. Constant division maintains vigorous stock – worth considering when old-fashioned double and semi-double primroses are grown. These are grown as choice border plants for a shady corner, not for bedding. They are very beautiful with their Victorian charm and much in demand for the popular cottage garden display that now seems so widely popular. But they have a bad reputation, establishing quickly and flowering freely for the first couple of years, but then dying out without warning. Lifting and dividing the plants every second year and planting them in fresh soil keeps them going.

All polyanthus and primroses enjoy a moist soil rich in organic matter with a little protection from the hot summer sun. They are versatile and beautiful plants for formal bedding schemes, but are equally useful in a herbaceous border or shrubbery. There is no definite role for these lovely plants and no need to lift them every year, unless you want the space they occupy for summer bedding.

The traditional role in which they have been used has become extended, so that they are now grown in a greater variety of situations. But wherever you put them their requirements are the same – moisture, coolness and a little shade.

BOG GARDEN PRIMULAS

The same could be said for these lovely plants, though their cultivation is markedly different. Bog garden primulas demand constant and consistent moisture, a few like the Himalayan cowslip, *P. florindae*, actually revelling in standing water. Though not essential, they do tend to grow better on an acid soil, but an alkaline one is perfectly adequate if plenty of organic matter is incorporated from the outset. A number of these primulas will tolerate full sun, but most prefer at least light shade, though they tend to become drawn in heavy, gloomy shade.

Most moisture-loving primulas are raised or sold in pots and can therefore be successfully planted at any time of year provided they are kept well watered. Spring is best, for the plants are then starting into active growth and establish much more quickly. If pot bound, the rootball should be slit with a knife as recommended for indoor plants. This allows the roots to escape quickly and ramify in the surrounding soil. Take out a hole with a trowel, large enough to take the rootball. Scatter a little fine peat around the roots, then firmly replace the soil.

Preparing the soil for moisture-loving primulas is not complicated, but before enthusing any more over their cultivation let us see what is involved. The medium advocated for polyanthus is ideal, though it must be kept much wetter. This is not too difficult for gardeners on heavy clay, except in very warm summers. To cover such even-

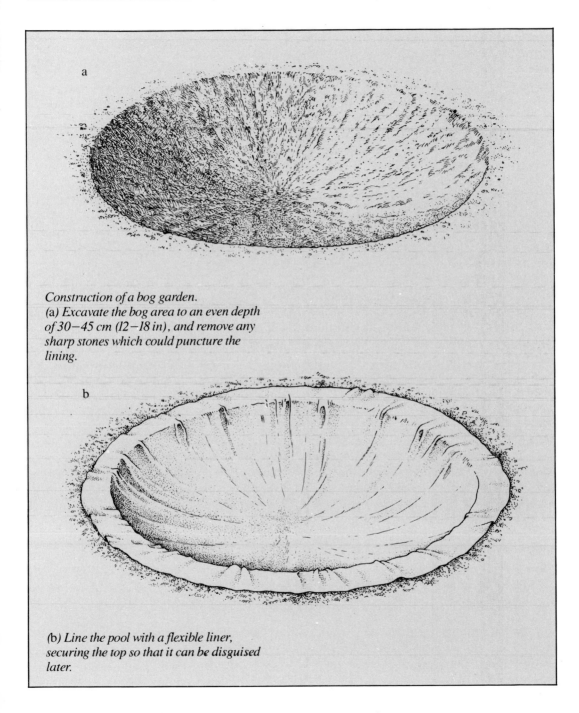

Construction of a bog garden.
(a) Excavate the bog area to an even depth
of 30 – 45 cm (12 – 18 in), and remove any
sharp stones which could puncture the
lining.

(b) Line the pool with a flexible liner,
securing the top so that it can be disguised
later.

(c) Fill with suitable moisture-retentive soil mixture, covering the overlap; water well and then plant. This stage shows the bog garden fully planted up.

tualities and provide for those who garden on free-draining soil, a special boggy area needs to be created.

This can be successfully constructed using a modern rubber or PVC pool liner. An artificial damp area can be just moister than the surrounding ground, or made into a proper bog garden. A true bog is preferable for most moisture-loving primulas. This looks most natural when attached to the garden pool, though it does not have to be. Construction is much easier if it is, however.

All that is required is a liner larger than necessary for the pool itself, the excess being spread out as a shallow pool about 30 cm (1 ft) deep. A retaining wall of loose bricks or sizeable stones is then laid to provide a barrier between

Opposite: *The oxlip,* Primula elatior, *is a great favourite for the wild garden and is seen at its best naturalized in grass.*

the pool and the bog area. This is filled with a peaty soil mixture (roughly equal parts by volume of soil and peat) over gravel. This provides a moisture-retentive medium, but allows excess water to drain from the roots. Water from the pool moistens the soil through the barrier, the soil surface being at least 5 cm (2 in) above mean water level.

If the wet area is to be in another part of the garden it can be provided by making a 45 cm (1½ ft) deep excavation and lining it with a pool liner. The top soil removed from the excavation is mixed with an equal amount of peat and

replaced. Once the soil has been put back it should be thoroughly soaked. Keep a close eye on it for drying out, until its behaviour is assessed. Some bog beds dry out much faster than others, without apparent reason – one problem that arises when the bog is separate from the pool.

The routine cultivation of bog garden primulas grown in specially contrived conditions presents few problems. The most important thing is to dead head the blossoms when they have faded. If they are allowed to set seed, hybrid seedlings could appear everywhere. These are not only likely to become inextricably mixed with the desirable plants, but growing in very wet conditions they are very difficult to weed out. If the flower heads are removed immediately they fade, however, this problem never has to be faced.

In the autumn, when the first sharp frost has killed off the leaves, remove all the debris. Rake it off gently with a spring-tined rake. If old foliage is allowed to remain, it provides a haven for over-wintering slugs and snails.

ROCK GARDEN PRIMULAS

Primulas grown on the rock garden need rather special conditions. All demand an open sunny position and a friable free-draining soil and very few popularly known as rock garden varieties can be grown elsewhere. If you are not planning a full-sized rock garden, an outcrop or similar feature will be necessary to grow them successfully. But preparation of both is very similar. You need to understand that whatever the scale of feature, primulas need a moraine for successful cultivation. In nature this is an accumulation of rocks and other debris found at the foot of and along the edges of glaciers. Not a heap of soil with rocks protruding from it, but a conglomeration of stones and debris. All the popular rock garden primulas will flourish in such conditions with a minimum of soil, provided there is constant moisture beneath.

Select stones to form the base of the feature and place them in position to create a retaining wall and define the outer contours of the formation. Make sure the strata are running the right way, as a stone placed upside-down or on its edge sticks out like a sore thumb and is virtually impossible to move once set in such a key position. Soft stones such as limestone and sandstone always need careful placing, whereas harder ones such as granite are almost totally devoid of clearly defined strata and can be used freely at almost any angle. However, it is important to use only one kind of stone – proper rock, not a collection of bricks and broken concrete – to create a natural looking feature. It must be built in the open. Once the key basal stones have been laid, fill between them with a mixture of two-thirds stone and one-third gritty soil. Old bricks and broken roofing tiles make an excellent filling where they are not likely to be exposed to view, if carefully camouflaged with a good depth of stony soil. Old tin cans and similar rubbish should not be used, for though they provide good drainage for a couple of years, they eventually rust away, causing subsidence. Having laid the base, filled the cavities, and formed a solid plateau, further stones can be lifted into position and the process repeated until the formation is of the desired shape and height. When completed, pockets can be excavated between the rocks to house primulas with special requirements and a suitable compost introduced. This is only necessary for the rarer, more tricky species.

Planting can be done at any time, but spring is best. The plants are just breaking into growth and with steadily improving weather conditions establish much more quickly than in autumn or winter. Summer planting is possible, but watering must be attended to regularly or the plants will suffer. Whenever they are planted, the plants' rootballs must be undisturbed. Only if a plant has become completely pot-bound is it necessary to break up the ball so that constricted roots can escape and spread into the surrounding soil. Always plant at the same level as the surface of the pot – never with the rootball protruding or it will dry out. On the other hand never bury the plant, so there is a risk it will rot off at the collar.

Once well established, rock garden primulas are easily managed and can be treated much as any other alpine plants. In late spring and early summer when they have finished flowering, remove their faded flower stems with a pair of shears or scissors to prevent them from setting seed and direct their energy into their sound establishment. Also cut away any dead or decaying foliage and

curtail the spread of vigorous varieties. Healthy outer pieces removed while tidying up can be potted as individual plantlets. At the same time of year rock garden primulas benefit from feeding. The safest fertilizer is bonemeal, which is dusted around each plant. Bonemeal is a reasonably mild slow-release natural fertiliser and is ideal, for unlike stronger quicker-acting kinds, it does not distort the character of the foliage. Never use general garden fertilizers like Growmore around growing plants as they are likely to scorch the foliage.

It is sometimes essential to water rock garden primulas in hot summer weather. A good watering once a week in prolonged dry weather is usually sufficient to ensure survival. Do not be tempted to sprinkle the plants with water each day as this encourages the roots to come close to the surface where they are even more vulnerable to drying out. Slugs are often a problem among rock garden primulas, so scatter slug pellets regularly to keep them under control. Slugs do not travel far so be sure to place the bait at reasonable spacings. Rain can spoil the pellets, so they should be protected with a piece of slate or tile propped up just enough for slugs to creep beneath, but not for birds to devour them.

ALPINE PRIMULAS IN POTS

The larger, more boisterous alpine primulas like *P. marginata* are quite at home in the open rock garden, even in a hostile northern climate. But if you wish to grow some of the more delicate-

looking, though still hardy kinds, to perfection, an alpine house or frame is the only place. There are little gems like *P. farinosa*, and to a lesser extent the fancy cultivars of *P. auricula*, which are also derived from an alpine species.

Most soil-based composts suit alpine primulas, John Innes No. 2 potting compost being the favourite. Cultivars of *P. auricula* like a little more organic matter added and a liberal quantity of sharp grit. Exhibitors have their own special concoctions which are a closely guarded secret, but for general cultivation and garden decoration about an extra 10% by volume of sedge peat and 20% by volume of sharp grit – flint grit used by poultry keepers is ideal – mixed into the standard John Innes potting compost No. 2 is quite satisfactory. Other alpine species and varieties are at home with a less richly organic medium. John Innes No. 2 potting compost with an extra 25% by volume of sharp grit is suitable for most, though adequate drainage in the bottom of the pot is vital. This can be the traditional crocks (broken clay plant pot), though a generous layer of coarse gravel would do.

When potting alpine primulas be careful not to over-pot. It is easy to provide a pot that is too large. The foliage can be misleading – vigorous and bulky, while the roots are quite modest in extent. Over-potting can be disastrous, for the compost quickly becomes stale and waterlogged, the roots rot and infestations of sciarid flies finish off the plant. It is better to keep primulas in tight pots and feed them regularly.

All alpine primulas benefit from feeding during the sumer. A pinch of fine bonemeal to the centre of each pot well watered in is the safest way to administer nutrients. Liquid feeds can be used, especially those formulated for tomatoes, but do not overdo them and cause uncharacteristic growth. Anyway liquid feeding should continue only from flowering time until mid-summer. Prolonged feeding produces lush plants which could succumb during the winter. Certainly fungus diseases like mildew are difficult to control when luxuriant foliage has developed after liberal feeding. Regard feeding as a supplement, for the compost will contain sufficient basic nutrients. Repotting is done biennially, if not annually, so there is unlikely to be any significant shortfall.

Apart from keeping an eye open for the inevitable aphids and vine weevils, alpine primulas need relatively little attention. It is different if you are exhibiting, of course, but for the average gardener routine cultivation means providing adequate water and ventilation. During summer and autumn the plants need not be kept inside, but benefit from a spell in the open. Plunge the pots in a frame or plunge bed and treat them very much as outdoor plants. Watering is critical for these alpines, but is unlikely to present any serious difficulty if the pots are plunged and the roots kept cool. Take precautions

Opposite: *Unquestionably the best known of all the Juliana hybrids is 'Wanda'. It is very widely grown and flowers over a long period in spring.*

against slugs, for these will appear in the best ordered garden. Scatter a few pellets among the pots every two or three weeks.

PRIMULAS IN THE PEAT GARDEN

A number of choice primulas revel in peat garden conditions. A peat garden is an increasingly popular feature which presents the gardener with opportunities denied him in the open ground. A full peat garden is not necessary, but a small feature with its typical virtues can easily be constructed in a shady corner. Ideally the structure is built with peat blocks, but on a small scale stone walling or bricks can be used, though they are not as aesthetically desirable. They also limit the scope for establishing rare primulas, like *P. bhutanica*, which are happier growing in a more or less horizontal fashion out of a peat block wall.

Whether you decide on a peat garden or small peat feature, the main ingredients necessary for success with peat-loving primulas are the same – a cool deep medium consisting of at least 60% good quality sedge peat and 30% clean neutral or acid soil with about 10% sharp grit added. This will retain moisture well, yet drain reasonably freely.

Peat-loving primulas like *P. cockburniana* are best planted in spring. They are usually grown in pots during their formative stages, so make sure that the root ball of peaty compost is well moistened before planting. A dry root ball will remain dry for a long time, even when planted in a moist friable medium. Do not break up root balls unless the roots are very congested, since most of this class of primula resent disturbance. Keep the new plants well watered for the first year and protect them from slugs, which revel in the cool moist

Construction of peat garden.
(a) Build blocks in the same manner as bricks in a wall, tying joints alternately. Use curved contours rather than acute ones.

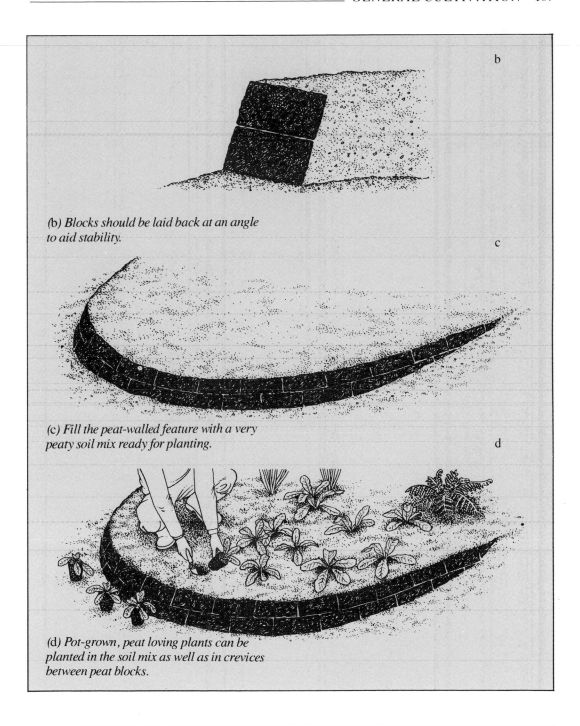

b

(b) Blocks should be laid back at an angle to aid stability.

c

(c) Fill the peat-walled feature with a very peaty soil mix ready for planting.

d

(d) Pot-grown, peat loving plants can be planted in the soil mix as well as in crevices between peat blocks.

conditions of a peat bed. Apart from removing faded flower spikes and occasionally dividing overcrowded plants, peat garden primulas look after themselves.

PESTS

☐ APHIDS

These common pests are popularly referred to as plant lice. There are many hundreds of species, many of which afflict primulas. The following aphids are generally thought to be the most devastating to primulas:

MOTTLED ARUM APHID
(*Aulacorthum circumflexum*)

This is a common pest of greenhouse grown primulas. Clusters of shiny yellowish or greenish aphids with distinctive dark markings infest stems, flowers, buds and foliage. Attacks are often followed by a growth of sooty mould which spreads across areas coated in the honeydew left by migrating aphids. It is also an important virus vector, which can carry any one of thirty or more plant viruses.

This aphid breeds parthenogenetically – that is, no sexual forms are known, the unfertilized eggs developing normally and producing adults directly. They breed all through the year, but are specially active during early spring, winged forms migrating regularly to infest other nearby plants.

Contact killers like malathion can be used, but a systemic insecticide based upon dimeothate is most effective. Spray regularly from early spring until late autumn.

PEACH POTATO APHID (*Myzus persicae*)

A common insect pest of primulas outside and under glass. Colonies congregate on the growing points, flower stalks and beneath the leaves of almost all primulas. Those growing under glass can be affected from early spring until the autumn. Outside it is rare to see any infestation before mid summer, but the rate of reproduction is rapid, so they can quickly reach epidemic proportions. Tell-tale signs are the twisting, curling and general distortion of leaves accompanied by honeydew and the eventual development of sooty mould. Like most other aphids this is an important virus vector.

These aphids, as their common name suggests, have an affinity with peaches and their near relations. The eggs overwinter on such trees, and in a frost-free greenhouse or home will continue to breed throughout the winter. The emergence of the first generation is largely influenced by the temperature, but it is usually on the wing by late spring. Most of the adults are greenish or pale yellow, but often pinkish forms are mixed among them.

If you have a likely overwintering host nearby you can break the life cycle by spraying this with a tar oil wash during the winter. Do this when the tree is completely dormant, during a dry, frost-free period. A thorough dousing will kill the eggs and delay any future infestation. Of all the aphids, this species has become most resistant to chemical controls. A spraying regime using a range of systemic and contact

aphicides is best, ringing the changes as often as possible and using the widest range available. There is no sure cure for these pests, but they can be kept under control by constant assault.

ROOT APHIDS (Many species)

There are many closely related root aphids of similar appearance which delight in attacking the succulent roots of primulas. They are common on primulas grown as house plants, but in the alpine house their effect is devastating if not checked. Often the first sign of trouble is the total collapse of a plant, which when knocked out of its pot has only vestiges of roots left surrounded by white, powdery, waxy aphids.

One of the most widespread primulas, P. veris, *flowers in the spring and grows happily when naturalized in grass.*

The life cycle of the species varies, but most reproduce parthenogenetically. Some live continuously on roots, others do so during one stage of their life cycle. All can cause extensive damage if not controlled. Drenching plants with malathion sufficient to penetrate the compost will have some effect, as will mixing diazinon granules into the compost. Neither is foolproof but should be worked into a repotting programme if infestation is extensive. Clean all plants of soil and wash out the pots with a mild disinfectant solution.

☐ MITES

BRYOBIA MITES (*Bryobia species*)

These are mites of the genus *Bryobia* which are not specific to primulas, but can cause damage similar to that resulting from red spider mite infestation. They are usually discovered in the open garden, attacks beginning in early spring and continuing until late summer. The leaves become discoloured, often with a bronze patchiness, eventually turning dry and crisp and dying prematurely. Unlike red spider mites, bryobia mites do not make webs. Instead of clustering under the leaves they tend to live on the surface. In autumn they migrate to a safe haven, like a fence or garden shed, over-wintering as eggs or adults.

Adult bryobia mites differ from red spider mites in having no bristles on their bodies and relatively long forelegs. They reproduce parthenogenetically, males being totally unknown. Some species reproduce rapidly, others slowly, but all are equally destructive. Spray regularly with malathion when bryobia mites are noticed.

GLASSHOUSE RED SPIDER MITE (*Tetranychus urticae*)

These tiny pests attack almost all plants grown under glass and are particularly troublesome where the atmosphere is dry. In a sun lounge, on a window ledge or in a greenhouse *P. obconica* regularly falls prey to these damaging pests. They occur to a lesser extent among alpine primulas grown in pots, though if these are stood outside from late spring onwards attacks should be minimal.

Infestations first appear as a fine light stippling on the foliage, sometimes developing into yellowish spots, particularly on the under-surfaces of the leaves. Close inspection with a magnifying glass will reveal colonies of brownish or reddish mites and clusters of minute rounded eggs. As the population of mites increases, as it can do at a prodigious rate, given warmth and a dry atmosphere, the foliage takes on a bronze or purplish hue, starts to twist and contort and becomes covered with a fine silky web. In severe cases plants can be almost completely defoliated.

The greatest problem occurs in late spring and early summer, when the female mites lay up to 100 eggs each on the undersides of the leaves. These are likely to hatch within two or three weeks, depending on the temperature. Reproduction ceases as autumn approaches, the adults seeking refuge in crannies in brickwork and similar dry places. They hibernate until the temperature rises in spring when they creep out and start reinfesting plants.

Chemical control of glasshouse red spider mites is rarely satisfactory, especially on primulas. Good glasshouse hygiene and a moderately damp atmosphere are the most suitable methods of control. If trouble persists, using derris (rotenone) will help, but this is not persistent and must be applied every few days. Recently a biological control has been developed using the predator, *Phytoseiulus persimilis*. This is really only helpful to commercial growers and unlikely to benefit the home gardener with just a few plants.

□ FLIES

GLASSHOUSE WHITEFLY
(*Trialeurodes vaporariorum*)

Probably the commonest pest of primulas indoors. A frequent assailant of both *P. malacoides* and *P. obconica*, disfiguring the foliage and making plants sickly. When a plant is disturbed myriad small white flies leap into the air but within a few moments settle back on the foliage. Where there has been a severe infestation sooty mould may develop.

Reproduction is mostly parthenogenetic and in a warm environment continues for most of the year. Adults live for three or four weeks, during that time laying up to 200 eggs. After a week or ten days these turn into nymphs, which scramble around the foliage until they find a suitable place to feed. The nymphs' legs and antennae degenerate and for a couple of weeks they become no more than feeding scales. They then enter a non-feeding pupal stage before emerging as adults.

There are a number of methods of chemical control, but all depend upon dogged persistence. Contact insecticides like malathion, pyrethrum and permethrin are effective, as are the systemic insecticides dimethoate and formothion. As with glasshouse red spider mite there is now a biological control, the parasite *Encarsia formosa*, but this is again impractical where only a few plants are grown. Vigilance and regular spraying or fumigating are the only reliable methods of control. However it must be said that greenhouse smokes used on *P. obconica* should be carefully monitored as their regular use can be a cause of damage to the foliage.

SCIARID FLIES (*Several species*)

These are an irritation rather than a major pest, though they can cause the collapse and death of pot-grown primulas.

Mainly inhabitants of waterlogged or stale soilless compost, these little flies are often seen creeping around the surface of the compost of plants that have not recently been repotted. They feed mainly upon fungal mycelium and decomposing organic matter, but in stale anaerobic conditions they often attack young root hairs, especially if these have been weakened or damaged by unsavoury compost conditions. A patch of decay will be the signal for the tiny worm-like larvae to attack, and this leads to further decay and further opportunity to attack. With a severe infestation, and stale compost, the plant will eventually succumb.

Adult sciarids are greatly attracted to soilless composts because of their rich organic content. They reproduce most of the year round in the home or under glass, the females laying up to 100 eggs at a time. After a week these hatch into myriad tiny whitish worms with brown or black heads. These feed for up to a month on decomposing peat and vulnerable hair roots in the compost before pupating in the soil.

Regular repotting to avoid stale compost conditions will ensure that sciarid damage is rare. If there is a bad infestation, drench the plants with malathion.

□ SLUGS (*Deroceras spp., arion spp., milax spp.*)

Slugs are common pests of primulas growing outside. They delight in devouring emerging shoots, often shearing off rock garden primulas at ground level. Most are active in the cool of the evening and revel in moist conditions. They hide during the day in accumulated garden debris. Good garden hygiene reduces their activities, but no matter how careful you are there will always be one or two around to cause trouble.

Apart from good garden hygiene and weed control there are chemical methods of control. These vary in their success, the traditional slug pellets being the most useful and easily administered. Liquid slug controls largely depend upon a good penetration of the soil, which on heavy clay is not easy. Pellets, on the other hand, are easily dispensed and have much improved in recent years. They used to be made with bran and metaldehyde and at the first sign of dampness in the air collapsed into a messy heap. Metaldehyde is still one of the commonly used ingredients, although some manufacturers are turning to methiocarb. The pellets now being produced have a greater damp tolerance. Science has given slug pellets a more protective coating, the best brands having a bluish colouration and being considerably more than showerproof.

Opposite: *A streamside provides an ideal home for yellow mimulus and moisture-loving candelabra primulas.*

Gardeners are naturally concerned about scattering slug pellets among their plants and rightly so. If birds or hedgehogs take a fancy to them, you will have a death on your hands, despite the manufacturers' claims that the pellets are deliberately made distasteful to birds and mammals. To avoid such a tragedy place the pellets close to the plants and protect them with a piece of tile or slate raised on a couple of sizeable stones. The slugs can then creep beneath and feed, but birds and mammals cannot get at them. Slugs are not great travellers, so place bait not more than 10 metres apart in any direction to ensure a comprehensive kill.

□ VINE WEEVIL (*Otiorrhynchus sulcatus*)

This is a most devastating pest of pot-grown primulas, and, in warmer areas, of those growing outdoors. There is rarely any evidence of its presence until plants collapse and die. An observant gardener may note a minimal amount of chewed foliage caused by the nocturnal feeding adults, but the soil-infesting larvae are the most destructive. These eat away most of the root system, usually killing the plant.

Adult beetles are small and brownish, and although rarely seen, as they are most active at night, look quite innocuous to the casual observer. They lay up to 1,000 eggs in the soil over a three or four month period. Most adults produce eggs as they are likely to be female, reproducing parthenogenetically in the absence of the rare males.

The eggs are laid close to the host plant and within a couple of weeks have hatched out and started to attack. They continue to feed for two or three months before pupating in cells in the compost. There is only one generation each year, but this is likely to be staggered, eggs, larvae and adults often being present at the same time.

Good hygiene will help prevent the appearance of vine weevils, but they can still appear in the best ordered greenhouse. Regular inspection of plants for signs of foliage damage and the occasional removal of plants from their pots to inspect the root balls for grubs are wise precautions. Incorporating HCH dust in the compost at potting time is a good deterrent, though this must be used sparingly. A mere dusting over the heap of compost to be used will be effective and is unlikely to harm the plants. Drenching the foliage of plants thought to be infested with a systemic insecticide does offer some measure of control as the insecticide is translocated around the plant and any part that is chewed is potentially lethal. This is not often recommended, but it can be useful if minor infestation is suspected and repotting is not desirable.

DISEASES

□ DOWNY MILDEW (*Peronospora spp.*)
This group of downy mildews affects a wide range of plants with varying degrees of severity. The foliage of infected plants often takes on a whitish hue with yellow or brownish spots and later a purplish furry mould. It is most serious on seedlings and recently divided plantlets and if not treated promptly can prove fatal. On adult plants it is a disfiguring nuisance. Badly infected plants should be destroyed, but if attacks can be caught in time, regular weekly spraying with zineb is effective.

□ LEAF SPOTS (*Ramularia primulina, Phyllosticta primulicola*)
The various leaf spots which attack primulas are disfiguring rather than debilitating. Polyanthus are most frequently affected, their leaves showing round yellow spots, which are often mouldy and surrounded by a crisp brown ring, which leads to degeneration of the tissue and results in it falling out to leave holes. Other primulas are differently affected, displaying brown or black spots. All leaf spots can be controlled by spraying regularly with benomyl or zineb.

□ ROOT AND FOOT ROTS (*Phytophthora spp., pythium spp.*)
Primulas, indoors and in the garden, are subject to various root and foot rots. These usually appear in over-wet growing conditions. The first symptoms are a general reduction in the size of new leaves, a yellowish colouration and what appears to be a wilt, despite the soil being moist. If the plants are flowering they are likely to be poorer than expected, the upper blossoms sometimes not opening properly and dying. In the final stages the plant collapses and the

upper leafy portion can be readily separated from the decomposing roots.

The fungal diseases which cause these rots are present in the soil or growing medium and can often survive long periods on decomposing organic matter. So where infected plants have grown, do not try further plants of the same genus for several years. The compost from pot grown diseased plants must also be carefully disposed of and the pots thoroughly cleaned in a disinfectant solution before re-use. Any plant showing signs of these maladies is best consigned to the bonfire.

□ PRIMULA RUST (*Puccinia primulae*)

Like most rusts, that which afflicts primulas does indeed look like rust – yellow or brownish rusty spots under the leaves. Sometimes found on primroses and allied species, it is not a seriously debilitating disease and so is rarely controlled chemically. Removing infected leaves to stop its spread is usually adequate.

□ VIRUSES

A number of viruses affect primulas, most transmitted by aphid vectors, although there are one or two soil-borne kinds. Cucumber mosaic virus is one of the commonest, causing yellowing mottling of the foliage and distortion and stunting of growth. Like all other virus diseases nothing can be done to control them and infected plants are best destroyed. In recent years nurserymen have been able to clean up some primula stocks using a micropropagation technique called meristem culture. The tiniest portion of growing point is taken and using sophisticated laboratory techniques enables the scientist to produce plants from virus-free growth cells.

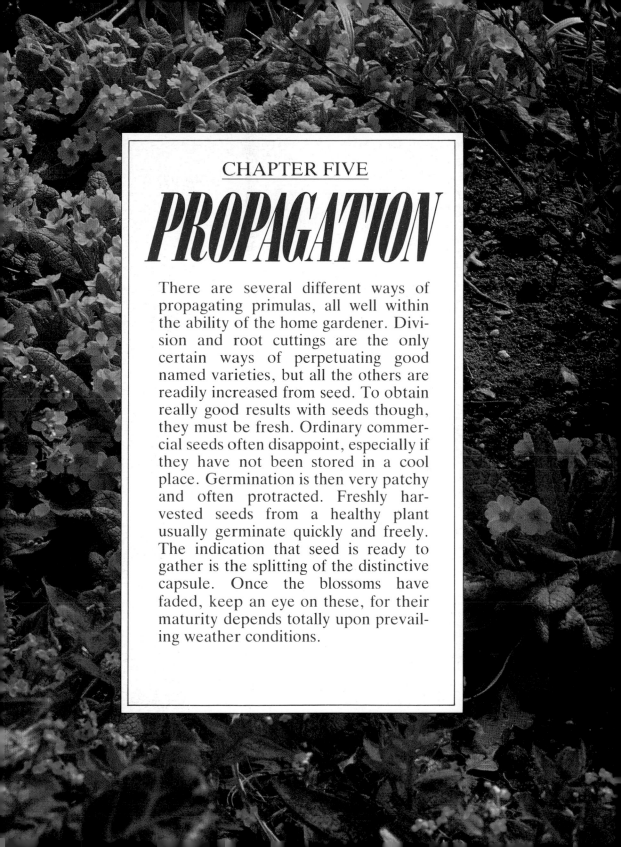

CHAPTER FIVE

PROPAGATION

There are several different ways of propagating primulas, all well within the ability of the home gardener. Division and root cuttings are the only certain ways of perpetuating good named varieties, but all the others are readily increased from seed. To obtain really good results with seeds though, they must be fresh. Ordinary commercial seeds often disappoint, especially if they have not been stored in a cool place. Germination is then very patchy and often protracted. Freshly harvested seeds from a healthy plant usually germinate quickly and freely. The indication that seed is ready to gather is the splitting of the distinctive capsule. Once the blossoms have faded, keep an eye on these, for their maturity depends totally upon prevailing weather conditions.

RAISING INDOOR PRIMULAS

These can be raised from seed sown under glass or on a windowsill at any time from mid-winter onwards. The ratio of light to warmth at the time of sowing must be in balance or seedlings will be sickly. It is much better to wait a week or two if conditions are not quite right and then sow, for though the plants may take longer to reach maturity, they will be much healthier and better balanced. Seedlings quickly appear from an early sowing in the warmth of a living room or kitchen, but in the poor daylight experienced in the early months of the year, even in a south-facing window, they can become drawn and leggy. The light and temperature ratio is so out of balance that the resulting plants are unlikely to make satisfactory progress.

Conversely, when plants are being raised in a greenhouse there is often over-optimism about the ability of the heating system to keep a stable temperature. When severe weather comes the temperature drops, causing a severe check to growth. This can retard the progress of the plants so that they lose two or three weeks, so it would have been better to delay sowing for a similar period and ensure continuous unretarded growth. A greenhouse does not need to freeze to cause problems: a mere drop in temperature at the critical seedling stage is enough.

All indoor primulas are raised from seed sown in trays or pans of good seed compost. Never be tempted to use ordinary garden soil for seed raising, even if it looks satisfactory, for it will be full of all kinds of pathogens and is likely to cause problems later on. Bought in primula seeds have only one aim in life – to germinate – so provide them with the best conditions you can. Never skimp on the cost of a tray full of good compost. Plants directly reflect the medium in which they are growing. It is vital to be selective about the composts you use. Proper seed composts have few nutrients in them but are perfect for germination. The shortage of nutrients ensures that the compost is unlikely to damage tender seedlings, and that the growth of moss and liverworts is restrained.

Soil-based composts like the John Innes range are suitable for most primulas, but quicker results and better young plants can often be raised in a good soilless compost. All peat soilless compost needs treating with a little reserve as it tends to have large air pockets within it. Fine primula seeds can easily get lost in spaces on the surface. So primula seeds will make a better start in a mixture of peat and fine sand. It is always desirable when selecting a seed compost to choose a well known branded kind rather than mix your own. The components of home made compost can be so variable, and the results so unexpected, that the little extra cash involved in purchasing properly mixed, scientifically balanced compost is an excellent investment.

Pans and trays are the usual containers in which to raise primulas. Fill these to within 1 cm (⅓in) of the rim with compost. If this is soil based, it should be firmed down and tamped level. But if the compost is soilless it must just be put into the tray or pan and tapped level – never firm it, as this excludes all the air, makes life awkward for the emerging seedlings, and is often difficult to wet, particularly if it has been allowed to dry out for an hour or two. With all composts, when filling a seed tray, firm the corners and the edges with your fingers to prevent the inevitable sinking there and the irritating prospect of all the seeds being washed to the sides where they germinate in a congested mass. Seed compost can be watered from above before sowing. This is particularly useful with soilless mediums, as they settle and undulations can be levelled with a pinch of compost before sowing takes place.

Sprinkle seeds thinly over the surface. A light covering of compost gently tamped and watered completes the job. Primula seeds are very fine, some looking almost like pepper, so they are quite difficult to handle. However, their distribution over the surface of the compost becomes easier if they are first mixed with a little silver sand. If this is poured into the seed packet, then shaken, the seeds are fairly evenly distributed throughout the sand, which can then be scattered with the seeds. Apart from acting as a carrier, the sand indicates the area over which the seed has been distributed. Once sown, the tray or pan can be stood in a bowl of water and moisture allowed to soak through. Watering from above, even from a watering can with a fine rose, can

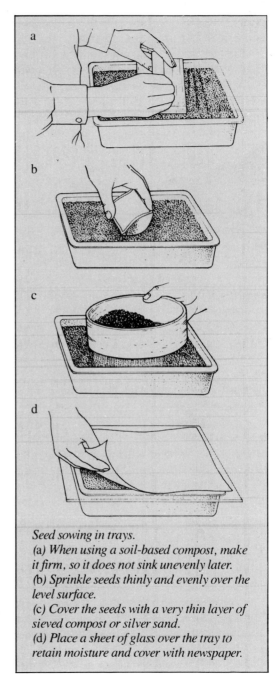

Seed sowing in trays.
(a) When using a soil-based compost, make it firm, so it does not sink unevenly later.
(b) Sprinkle seeds thinly and evenly over the level surface.
(c) Cover the seeds with a very thin layer of sieved compost or silver sand.
(d) Place a sheet of glass over the tray to retain moisture and cover with newspaper.

redistribute or disturb fine primula seeds and impair germination.

Most primulas raised indoors benefit from being stood over a heating cable, as warm compost promotes rapid germination. The combination of a soil-warming cable and a sheet of glass over the pan or tray can create a surprisingly effective micro-climate. Similarly, a sheet of newspaper placed lightly over a seed tray will act as perfect insulation yet still allow sufficient light to pass through. It is important to remove both glass and newspaper immediately germination takes place. Once the seedlings are emerging maximum light is vital to ensure that they develop into stocky plants. After germination primula seedlings sometimes keel over with damping off disease. This is encouraged by the close atmosphere created by emerging seedlings that have been sown too thickly, or even by too warm and humid conditions. Prevention is better than cure, so immediately seedlings have emerged give them a routine watering with Cheshunt Compound or a fungicide based on benomyl. It is wise to then repeat this treatment every ten days or so until the plants are well established.

Once primula seedlings are large enough to handle they should be pricked out. Lift and transplant them into trays so that they can develop as individuals. Most standard seed trays accommodate 35 plants. Ideally the seedlings should have their seed leaves or cotyledons fully expanded and the first rough leaf in evidence. Great care should be taken in handling them as

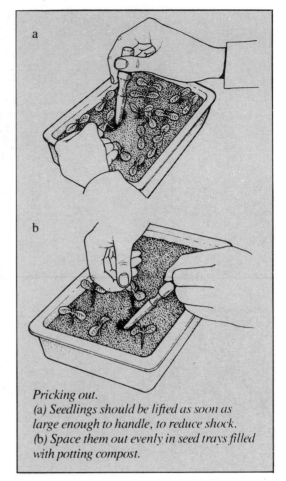

Pricking out.
(a) Seedlings should be lifted as soon as large enough to handle, to reduce shock.
(b) Space them out evenly in seed trays filled with potting compost.

they are very delicate and brittle. Never hold a seedling by its root or stem, always take hold of it by the seed leaf. Plant each seedling slightly lower than it was in its original tray or pan, indeed it can be planted with its seed leaves at compost level if it is short-jointed, strong and healthy. Deep planting will rarely turn a drawn, etiolated plant into a short healthy one. Such a seedling is more likely to rot off and die.

Gold laced polyanthus are beautiful old florists' varieties with a dark ground colour and gold lacing on the edges of the petals.

If seedlings are pricked out into a potting compost they are unlikely to need feeding before being potted. Little special care is required while the young plants are developing, provided they have plenty of light, are watered regularly and are watched for pests and diseases. The occasional use of a systemic insecticide and a systemic fungicide, if necessary, will keep greenfly and mildew at bay.

PROPAGATING BOG GARDEN PRIMULAS

Bog garden primulas of the candelabra type and other moisture lovers such as the drumstick primula, *P. denticulata*, can be raised from seeds in exactly the same way as the indoor species. Packeted commercial seeds are best sown during early spring, but where there is access to freshly gathered seed, mid to late summer is the period for sowing. The only variation in the growing technique is to keep as cool a temperature as possible. The use of heating cables or a propagator is undesirable. A well ordered cold frame is much to be preferred.

Once primulas of this kind are well established in seed trays or small pots they should be hardened off before planting out. The plants are weaned from a protected environment and prepared for life in the open garden. A cold frame is the ideal place for this for in cold weather the frame light can remain on, whereas if the weather turns warm it can be removed entirely. A gradual transition over the course of two or three weeks is usually necessary.

First of all the frame light is lifted to give ventilation. This is gradually increased until the frame light can be removed entirely during the day. It is then lifted slightly at night to allow ventilation, except when frost threatens. Eventually the light is left off night and day. If you do not have a frame and your plants have been raised in your house or sun lounge, the same effect can be provided by taking the plants out during the day and standing them in a sheltered place, then returning them to the house for the night. Weaning can continue in just the same way until the plants look hardy, as indicated by stiff foliage, often of darker green. If the plants turn bluish-green, weaning has

been too swift and the plants have been checked, but they will eventually grow out of it.

RAISING ALPINE PRIMULAS

The seeds of alpine primulas in particular benefit from being sown when freshly gathered. For many gardeners this is impossible because they depend upon seeds sold by seed companies or distributed by specialist societies like the Alpine Garden Society. Obviously these organisations have to gather, clean, package and distribute seeds, so the seeds are several months old when received. Fresh seeds are only available from one's own plants or a friend's. When alpine gems like *Primula frondosa* and *P. rosea* are to be grown, it is well worth securing some recently gathered seeds from a fellow enthusiast.

When fresh seeds have been harvested they should be sown in pans of gritty seed compost as quickly as possible and plunged into sand in a cold frame. Before sowing, the compost must be thoroughly soaked by standing the pans in a bowl of water. No more than gentle watering from above will then be necessary until germination takes place. Freshly gathered seeds are always sown during the summer, so frame lights must not be used as they will create a hothouse effect which is both unnatural and damaging to emerging seedlings. Instead, cover the frame with wooden laths or twiggy branches to create a cool airy environment.

Packeted primula seeds used always

to be sown during the summer, which saved the gardener time and space during the busy spring season. But it produced plants with a tendency to soft growth towards late summer. Unless great care was taken with over-wintering, or feeding to induce hard growth, losses were likely to occur during severe weather. The seed was also much older and less viable than that sown during the spring. Spring sowing is now much preferred, for the seeds are more likely to germinate freely, and the plants can be grown alongside summer-flowering annuals in a cold frame or greenhouse.

Most alpine primula seeds, especially those from high alpine slopes, will not germinate satisfactorily until they have experienced a winter. This can be contrived artificially by sowing them in the normal way, then freezing them in a domestic freezer. If the seeds are received soon enough they can be subjected to a natural winter. The only disadvantage to allowing pans to stand unprotected is that birds and mice will often scratch in the compost and disturb the seeds. But a covering of fine mesh wire netting will overcome this problem.

As soon as the seedlings appear and are growing away strongly they should be pricked out into pans or seed trays. Use a gritty compost once again, with up to 25% by volume of sharp grit and be sure to add sufficient crocks or gravel to the bottom of the pans or trays to guarantee free drainage. When the seedlings are well established move them into individual pots before planting out.

INCREASING PLANTS BY DIVISION

Many of the hardy perennial primulas are easily increased by division. This is the most useful way of perpetuating good colour forms of candelabra and drumstick kinds as well as polyanthus and primroses. Auriculas are also propagated this way, except commercially where micro-propagation is now widely used to bulk up plants quickly and to rid them of virus. Several of the fine named alpine primulas, especially those derived from *P. marginata*, also increase freely from division. A number of them have to be divided regularly to maintain their vigour.

Division is best undertaken immediately after the plant has flowered, though some gardeners divide bog primulas in the early spring just as new growth is showing. Polyanthus, primroses and spring-flowering alpine primulas must be divided after flowering. Lift the clumps and retain the healthiest young growths. Discard hard woody growth from the centre of the plant, as this never regenerates satisfactorily. Some alpine primulas grow from what appear to be extended stems along which roots are produced. Treat these more or less as cuttings and remove them, each with a little root attached, then plant them like ordinary divisions.

When primulas are lifted for division, it is usually quite clear how to divide them, for unlike many other perennials they separate clearly into individuals. There is rarely any need to take a knife

Divide old, overgrown clumps of primulas by digging up and dividing into smaller clumps with a handfork. Discard exhausted pieces and replant strong young ones.

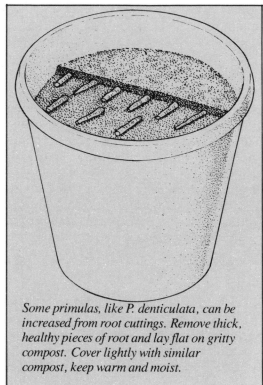

Some primulas, like P. denticulata, can be increased from root cuttings. Remove thick, healthy pieces of root and lay flat on gritty compost. Cover lightly with similar compost, keep warm and moist.

to them – just pull them apart. The larger coarser kinds can be planted directly into their permanent positions or lined out in nursery rows until their planting sites are ready. Polyanthus used for spring bedding are often lifted annually after flowering to make way for summer-flowering bedding. They are divided and planted in a shady part of the garden until the autumn, when they are put out again for the spring bedding display. The alpine kinds are less resilient and should be potted individually and placed in a cold frame until established. They can then be planted out. The same applies to any rare or unusual kinds where every piece of divisible plant is needed. Divisions that have only a tenuous hold on life should be potted and nurtured separately.

TAKING ROOT CUTTINGS

Root cuttings are usually taken from herbaceous plants like oriental poppies, globe thistles and sea hollies, but a number of primulas can be increased in this way too. The benefits are the same as for division, for the plants that result are exactly the same as the parents, yet because only small sections of the root are needed, the parent plants yield many more progeny. Most candelabra primulas can be propagated this way, but it is most widely used with good colour selections of the drumstick primula, *P. denticulata*.

Root cuttings are taken during the dormant period, the adult plant being lifted carefully to expose the roots. Suitable cutting material is removed and the plant replanted without harm. It quickly re-establishes and usually flowers as if it had not been disturbed. The best roots to use for cuttings are the young, vigorous ones. Choose those which look strong enough to survive detachment from the parent plant without desiccating, yet not be coarse and woody.

Prepare cuttings from lengths of root, each no more than 3 cm (1¼ in) long, and lay them horizontally in trays of an equal parts mixture by volume of peat and sharp sand. Then lightly cover and water, and place in a cold frame. The sections of root should sprout in the spring and give rise to small independent plants. Once these have a small root system they can be potted individually and grown on until firmly established.

APPENDIX

SOCIETIES

□ THE NATIONAL PRIMULA AND AURICULA SOCIETY

There is a number of societies that are relevant to the growers of primulas. The most important is the National Auricula and Primula Society. Founded some 115 years ago as the National Auricula Society it was orientated solely towards the preservation, perpetuation and maintenance of the florist's auricula. Since then it has come a long way, so that it embraces almost every aspect of the genus *Primula*. There are three regional sections – the Northern Section, Midland Section and Southern Section. All are united in their endeavours to further interest in primulas by holding shows and producing an annual year book.

Northern Section	Midland Section
D. G. Hadfield,	H. A. Cohen,
146 Queens Road,	'Shenley',
Cheadle Hulme,	9 Reddings Road,
Cheshire	Moseley,
SK8 5HY.	Birmingham 13.

Southern Section
L. E. Wigley,
67 Warnham Court Road,
Carshalton Beeches,
Surrey.

□ THE AMERICAN PRIMROSE SOCIETY

While directed towards primroses, information and publications are produced about other primulas, the term primrose being rather loosely interpreted. Contact: Brian Skidmore, 6730 West Mercerway, Mercer Island, MA 98040, U.S.A.

□ ALPINE GARDEN SOCIETY

Although having a wide remit and covering almost all plants that can be loosely termed alpines, this society does have a strong contingent of primula lovers. It produces a quarterly bulletin, publishes other serious works, has a library service as well as an international seed distribution scheme. Groups meet and hold shows throughout the British Isles. There are over 7,000 members in some 30 countries worldwide. Contact: E. M. Upward, Lye End Link, St. John's, Woking, Surrey GU21 1SW.

□ SCOTTISH ROCK GARDEN CLUB

A very similar organization to the Alpine Garden Society with its own publications and regular shows and meetings. Primulas figure highly among its members' interests. Contact: D. J. Donald, 'Morea', Main Road, Balbeggie, Perth, PH2 6EZ.

INDEX